Good

Measures

A Practice Book to Accompany

RULES OF THUMB

Good Measures

A Practice Book to Accompany

RULES OF THUMB

FOURTH EDITION

Jay Silverman

Elaine Hughes

Diana Roberts Wienbroer

 **McGraw-Hill College**

Boston   Burr Ridge, IL   Dubuque, IA   Madison, WI
New York   San Francisco   St. Louis   Bangkok   Bogotá
Caracas   Lisbon   London   Madrid   Mexico City   Milan
New Delhi   Seoul   Singapore   Sydney   Taipei   Toronto

McGraw-Hill College

A Division of The McGraw·Hill Companies

ISBN: 0-07-092075-3

Editorial director: *Phillip A. Butcher*

Sponsoring editor: *Tim Julet*

Marketing manager: *Lesley Denton*

Project manager: *Nancy Martin*

Production supervisor: *Michael R. McCormick*

Designer: *Kiera Cunningham*

Supplement coordinator: *Nancy Martin*

Composition: *Shepherd Inc.*

Typeface: *11/13 Bookman Old Style*

Printer: *R.R. Donnelley & Sons Company*

http://www.mhhe.com/writers

About the Authors

A graduate of Amherst College and the University of Virginia, Jay Silverman has received fellowships from the Fulbright-Hayes Foundation, the Andrew Mellon Foundation, and the National Endowment for the Humanities. Dr. Silverman has taught at Virginia Highlands Community College and at Nassau Community College where he received the Honors Program Award for Excellence in Teaching and where he also teaches in the College Bound Program of the Nassau County Mental Health Association.

Elaine Hughes moved to New York City from Mississippi in 1979 to attend a National Endowment for the Humanities seminar at Columbia University. She has taught writing for more than twenty-five years, primarily at Hinds Community College in Raymond, Mississippi, and at Nassau Community College. Since her retirement from NCC and her return to Mississippi, she has conducted many writing workshops for the Esalen Institute and for other organizations. She is also the author of *Writing from the Inner Self*.

As Chair of the English Department of Nassau Community College for six years, Diana Roberts Wienbroer coordinated a department of 150 faculty members and served on the Executive Council of the Association of Departments of English. Besides teaching writing for over thirty years, both in Texas and New York, she has studied and taught film criticism. She is also the author of *The McGraw-Hill Guide to Electronic Research and Documentation*, 1997.

The authors have also written *Rules of Thumb and Rules of Thumb for Research*, both available from McGraw-Hill.

Contents

PART I: CORRECTNESS

Part II: PUTTING A PAPER TOGETHER

PART III: MEETING SPECIFIC ASSIGNMENTS

PART IV: WRITING WITH ELEGANCE

To the Student

We have written *Good Measures* to give you practice in applying the concepts presented in *Rules of Thumb: A Guide for Writers*. At the top of each page, we've put the corresponding page numbers from *Rules of Thumb* so that you can look up the relevant rule.

Good Measures presents two kinds of exercises. Some are drills that offer a quick way to test your understanding. But drills can take you only so far; the real proof of your knowledge comes in actual writing. Therefore, we've also included writing activities which require you to apply your knowledge. In some cases, the required writing is very brief—making up single sentences. But often we've given longer assignments that we hope will be enjoyable as well as instructive. The drills in *Good Measures* give you ways to "measure" your learning by testing yourself. The writing activities present "measures" you can take to practice new skills and to grow as a writer.

Jay Silverman

Elaine Hughes

Diana Roberts Wienbroer

To the Instructor

We are eager for your reactions to *Good Measures*. We would like to know if it meets your needs in the classroom; we also welcome your suggestions for improvements.

Please send any comments on *Good Measures*—or on *Rules of Thumb*—to us directly:

Jay Silverman, Elaine Hughes, Diana Roberts Wienbroer
Department of English
Nassau Community College
Garden City, New York 11530-6793

Acknowledgments

We particularly wish to thank two people for their enthusiasm and support for *Good Measures:* Beverly Jensen, whose encouragement and ideas contributed to every page; and Laurie PiSierra, whose expert editor's eye guided the quality of the book.

Special thanks go to our colleagues—Jeanne Hunter and Hedda Marcus, at Nassau Community College, and Susan Finlayson, at Adirondack Community College, who gave us sound advice on content and audience.

We are also grateful for careful reviews, accompanied by specific suggestions for improvements by Robin Calitri, Merced College; Bill Nagle, Middlesex Community College; and Nell Ann Pickett, Hinds Community College.

Our thanks also go to John Fitzgibbon, Jane Malmo, and Ernest Migliaccio for their creative ideas; to Kiera Cunningham and Joan Greenfield for their inviting designs; and to Lesley Denton, Alison Husting, and Tim Julet at McGraw-Hill for their consistent support of the project.

We are especially grateful to our students who provided responses to the exercises and helped us to shape the book, and in particular to Jennifer Franceschi-Wood for permission to reprint her research paper.

Jay Silverman

Elaine Hughes

Diana Roberts Wienbroer

P A R T   I

Correctness

Confusing Words

AFFECT / EFFECT, ITS / IT'S

In the following sentences, cross out any incorrect words and write corrections above them.

Example: The trade deficit continues to ~~effect~~ *affect* the economy.

Affect / Effect

1. The special affects made the whole production magical.

2. A hot toddy can affect how you feel when you have a bad cold.

3. After an MRI (magnetic resonance imaging), the magnets have no lasting effect on the patient.

4. The tall man in the second row effected the view of the entire audience.

5. A household fire effects the whole family, and the effects last for years.

Its / It's

1. Even if its July, its still best to pack a sweater for a trip to Maine.

2. Ingrid Bergman had only a minute to return the key to it's proper place before the villain would notice its absence.

3. The snake vibrates it's rattle when it senses that it's in danger.

4. Its a miracle that no one was hurt when the car rolled over on it's side.

5. When its noon in New York, its six o'clock in Paris.

LAY / LIE, PASSED / PAST

In the following sentences, cross out any incorrect words and write corrections above them.

lying
Example: The clue to the corpse's identity was ~~laying~~ in the bushes.

Lay / Lie

1. Cleopatra laid on the divan, eating grapes; as she got up, she laid the grape stems on the cushion.

2. Once the foundation was in place, they began laying the floors, but the manager continued lying about the date of occupancy.

3. Every night while I was laying in bed trying to sleep, the smell from the paper factory kept me awake.

4. The archivist laid each picture flat in the drawer, but the last one was laying off-center.

5. The patient lay comatose while the nurses quickly prepared the instruments for surgery.

Passed / Past

1. The train sped passed the platform of waiting passengers.

2. The cat has never eaten tuna in the past because we had always bought the kind that was packed in oil.

3. Kip past the ball behind his back to set up Nancy, who made a slam dunk.

4. Several years ago, the total of the human population past five billion.

5. In crossbreeding tulips, growers hope that the most attractive traits from the past can be genetically past on.

THAN / THEN, THEIR / THERE / THEY'RE

In the following sentences, cross out any incorrect words and write corrections above them.

then
Example: Chop all the ingredients; ~~than~~ heat the wok.

Than / Then

1. Beethoven's late quartets are more extraordinary then his earlier, more popular music.

2. Advertisements urge you to buy whatever you want and then pay later.

3. The shares in L.A.Gear are worth 25 percent less today then they were last week.

4. In the dry season, this canal clogs up now and than, diminishing the water supply.

5. The chiefs in the southern part of Cameroon exercise much less power then the chiefs in the west.

Their / There / They're

1. Every year in August, their is a Twins Day Festival in Twinsburg,

   Ohio.

2. In *The Brothers Karamazov*, all of the brothers resent there

   father's silliness.

3. We turned the corner and there were my aunt and uncle standing

   in the middle of the dusty road next to there stalled car.

4. When overtime work is available, their always the ones who show

   up first.

5. There the only people I know who would rather read about a place

   than go there.

TO / TOO / TWO, YOUR / YOU'RE

In the following sentences, cross out any incorrect words and write corrections above them.

lo

Example: Employers are required by law ~~too~~ scrutinize the investments included in pension plans.

To / Too / Two

1. In order to snorkel safely, you need to have the proper equipment.

2. To many is as bad as to few.

3. Malcolm's hair is to thin to wear combed over the top of his head.

4. Evidence has recently come to light that to much passive smoke is harmful to one's health.

5. Whitman's subject matter was to risqué, for many readers of his day—and ours to.

Your / You're

1. Your lucky that you've never been in Uncle Eddie's fish store on a hot day.

2. When your cue comes, act as if your not sure which way to walk.

3. You're first step is to wash your hands with an antibacterial solution.

4. When your in a foreign country, your sense of time is altered.

5. You're going to be surprised when you start to record you're sugar consumption—one cup of catsup may contain seventeen teaspoons of sugar.

**A / AN, ACCEPT / EXCEPT, ETC., GOOD / WELL,
LOOSE / LOSE, KNEW / KNOW / NEW / NO / NOW,
HAVE / OF, QUIET / QUIT / QUITE,
WERE / WE'RE / WHERE, WEATHER / WHETHER,
WHO'S / WHOSE, WOMAN / WOMEN**

In the following sentences, cross out any incorrect words and write corrections above them.

women
Example: I don't believe that any ~~woman~~ were interviewed for the position of comptroller.

1. A room full of three-year-olds will never be quite.

2. Ara decided to except the director's criticism except for the comment about her favorite yellow hat.

3. After a night of driving through the jungle, the jeep had three lose wheels and the driver didn't even now it.

4. If the weather is favorable, were climbing to the top of Arthur's Seat when we return to Edinburgh.

5. During a play, the performers know who's going to enter next, based on whose cue has been spoken.

6. Trevor would have accepted the duchess's invitation, accept that he had already decided to elope that night.

7. Sally Ride, the first women in space, has since served as a presidential adviser on scientific research.

8. It was never a question of whether he would be a doctor, but weather he would be an orthopedist or a pediatrician.

9. Costello thought that Abbott was asking "Whose on first?" when Abbott was actually telling him about a first baseman named Who.

10. In the desert, were the temperatures rise and drop suddenly, you would have to wear extra layers of clothing.

11. The trucker has to drive quiet a long way before he is able to quite for the night.

12. The uncontrolled forest fire might of destroyed their house, their garage, their car, etc.

13. There has been much discussion of late as to whether a women will be chosen as the next ambassador.

14. In typical scenes from Chekhov, the characters know that now is the time to speak their true feelings or they will forever loose their chance—yet they say nothing.

15. During World War II, the military would not except a African American serving in the same unit as white Americans.

16. They should of known that there would be problems bringing so much luggage onto the plane.

17. The loose rock debris, called "scree," makes it very easy to loose one's footing.

18. Loons are quiet good swimmers and fliers, but they cannot walk very good.

19. Because of the knew code strip in one-hundred-dollar bills, it's harder know to pass counterfeit bills than it used to be.

20. The doughnut was originally called "dow-knot," a English treat made each year for Christmas.

CUMULATIVE EXERCISE

In the following paragraph, cross out any incorrect words and write corrections above them. For example, the first line should be corrected to read:

too
Many readers find the last part of *Huckleberry Finn* to be ~~to~~ drawn out.

Many readers find the last part of *Huckleberry Finn* to be to drawn

out. There disappointed by the way Tom Sawyer dominates Huck.

They're also uncomfortable when Tom makes Jim, the runaway

slave, go through childish games in order to win the freedom he

should of had in the first place. While Jim is laying chained up in a   5

tiny cabin, Tom makes an elaborate but needless plan of escape.

The plan seems to take forever. Once readers have past the

excitement of Tom's arrival, the book looses its charm. And yet, their

disappointment doesn't ruin their experience of the novel. Its far

less important to them then the joyful effect of most of the book.   10

CUMULATIVE EXERCISE

In the following paragraph, cross out any incorrect words and write corrections above them. For example, the first line should be corrected to read:

You're
The serviceman casually told Ilene, "~~Your~~ going to have to buy a

The serviceman casually told Ilene, "Your going to have to buy a new   15

hard drive for this computer. Your present one is too dirty." This was

not good news. His statement effected Ilene in several ways: her

mouth got to dry to speak; she felt the need to lay down; and the

room dimmed so that she couldn't see very good. She thought to   5

herself, "Where are we going to find the money to pay for this? We're

a nonprofit organization. Its going to cost us a lot of money. The

expiration date on the warranty has now past. Or has it?" Their was

a glimmer of hope. She remembered something. "They're not going

to like this, but it was their idea in the first place." Ilene searched   10

through her desk for the service contract. She found it, and there,

on the bottom line, was the following sentence: "If you purchase this

machine in the month of February, then you receive two additional

months of free service." This was more then she could have hoped for. They had purchased the machine last February 29th—the Leap

Year Special. Ilene congratulated herself for being one smart

women!

Writing Activity

Select any word from column A, and use it in a sentence about any topic in column B. Be sure to use all the words and all the topics.

For example, if you choose *there* and *something that leaks*, you might write

There is nothing as maddening as a dripping faucet at midnight.

Column A	*Column B*
you're	something that tickles
its	something that is clogged
too	something that leaks
than	something that smells wonderful
there	something that stinks
affect	something that burns
their	something that turns over

One Word or Two?

In the following paragraphs, look for two words where there should be one, or one word where there should be two. Cross out any incorrect words and write the correct words above them. For example, the first line should be corrected to read:

Anyone
~~Any one~~ who has ever seen an exhibit of early American quilts can

Any one who has ever seen an exhibit of early American quilts can

understand why quilts, nowadays, are being recognized as an

important form of American art. Some how the women of early

America—for it was the women who made quilts—perfected the

chore into an art form. During the years when the United States was   5

first being settled, everybody made quilts out of necessity. In fact,

they were an early form of recycling. Worn clothing was torn a part

and cut into alot of different-sized pieces. These scraps were then

sewn together randomly to make covers and backs. Quilts made this

way were called "crazy" quilts. Sometimes they were stuffed with   10

straw, sometimes with corn husks. Through out the harsh winters

the quilts helped to keep the settlers warm.

In fact, the haphazard piecing of odd scraps was only the beginning.

Inspiteof the crude materials that the women often had to work

with, the cutting and piecing of the scraps of fabric became   15

increasingly intricate. Nevertheless, with out the benefit of sewing

machines these women created beautiful hand-sewn quilts in order

to help both themselves and eachother. Nobody could deny that the

quilts are works of art.

Spelling

EI / IE

Insert either *ei* or *ie* into the spaces in the words below.

> ***Example:*** n___ghbor   n*ei*_ghbor

1. dec____ve

2. bel____ve

3. fr____nd

4. fr____ght

5. p____ce

6. n____ce

7. c____ling

8. s____ze

9. anc____nt

10. w____rd

11. br____f

12. th____r

13. rec____ve

14. v____n

15. l____sure

16. w____gh

17. conc____ted

18. rec____pt

19. for____gn

20. sl____gh

Spelling

T OR *TT*? *P* OR *PP*?

In each of the words below, fill in the blank with one or two of the letter indicated.

    Example: dro___ing   dro*pp*ing

1. wri____er	(t)		11. dra____ed	(g)
2. occa____ion	(s)		12. cra____ed	(m)
3. begi____ing	(n)		13. hi____ing	(d)
4. sto____ed	(p)		14. ru____ing	(l)
5. occu____ed	(r)		15. sho____ed	(p)
6. whi____ed	(p)		16. bla____ing	(m)
7. wi____ed	(p)		17. hi____ing	(t)
8. jo____ing	(g)		18. wri____en	(t)
9. qui____ing	(t)		19. dro____ing	(p)
10. refe____ed	(r)		20. ro____ery	(b)

20

Spelling

WORD ENDINGS

In the following paragraph, cross out any incorrect words and write corrections above them. For example, the first line should be corrected to read:

Coming
~~Comming~~ home with a new baby can be a trying time as well as a

Comming home with a new baby can be a trying time as well as a

joyous one. In the begining parents may not be prepared for the   15

intensity of the baby's crying. Psychologist confirm that crying

causes much frustration in new families. The baby's cries can have

many meanings: "I need food"; "I need to be held"; "I need to be   5

changed." But when feeding, holding, and changing don't work,

parents don't know what they are suppose to do, and they feel like

crying themselves. Gradually, however, the parents learn to interpret

different cries; gradually they learn tricks for relieving discomforts

(like walking or dancing while holding the baby close) and thus   10

stoping the cries; and gradualy the baby becomes more use to the

inner and outer discomforts of daily life. The biggest danger for

familys at these times is that frustration might spill over into fighting between parents or into child abuse. New parents need to know that their frustration is normal and that within three months things will be better. Instead of takeing their anger out on each other or on the baby, they should help each other through the trying times. Latter they may look back with pride, fondness, and even humor at the adventure of their first three months of parenthood.

Spelling

PREFIXES AND SUFFIXES

In the lists below, underline the misspelled word and write the correct spelling in the space to the right. (Only one word in each line is misspelled.)

> *Example:* postscript pastime <u>partime</u> halftime   *part-time* _____

1. disappear disatisfaction disappoint disillusion   _____

2. unsatisfactory uninterested unecessary unnoticed

            _____

3. hopful careful really lonely   _____

4. environment government committment disappointment

            _____

5. totally truely hopefully finally   _____

6. debatable incredible sensable noticeable   _____

7. arguement appointment suddenness carefulness

            _____

8. fourty ninety fourth nineteenth   _____

9. caring arguing loving debateing   _____

10. tarring dragging gunning begining   _____

Spelling

TRICKY WORDS

In the lists below, underline the misspelled word and
write the correct spelling in the space to the right.
(Only one word in each line is misspelled.)

Example: physical <u>psychatrist</u> psychology physiology

psychiatrist

1. intresting armrest interest arresting   _____

2. definitively infinitely definately defiantly   _____

3. repetition seperate particularly familiar   _____

4. profession personnel paralell professor   _____

5. absence sense scene lisence   _____

6. rhythm vacuum amateur athelete   _____

7. usually necessary opinion probaly   _____

8. surprise exaggerate embarass accommodate   _____

9. thoroughly actually realy immediately   _____

10. marriage opinon similar circular   _____

Spelling

CUMULATIVE EXERCISE

In the lists below, underline the misspelled word and write the correct spelling in the space to the right. (Only one word in each line is misspelled.)

 Example thorough through <u>thorougly</u> roughly *thoroughly*

1. accommodate mattress marrage hurricane _____

2. ancient chief weird sieze _____

3. sking beginning running marrying _____

4. occurred prefered referred murdered _____

5. committed committee committment admitted _____

6. probably repetition intrest eliminate _____

7. lovable enviroment management sensible _____

8. Wednesday Febuary business opinion _____

9. arguing arguement separation debating _____

10. monkeys attorneys tries vallies _____

11. singular similar familar particular _____

12. drugged dragged dropped interferred   _____

13. believe conceited receive piece   _____

14. truly really succesfully lonely   _____

15. hopeful noticeable ninty careful   _____

16. dissatisfaction unnoticed dissappointment misspell

_____

17. embarrass necessary usually occasion   _____

18. writer quitting writting quitter   _____

19. restlessness thinness occurence openness   _____

20. definitely seperate criticism ridiculous   _____

Spelling

CUMULATIVE EXERCISE

In the following paragraphs, cross out any incorrect words and write corrections above them. For example, the first line should be corrected to read:

definitely

Earning a Ph.D. is ~~definately~~ a demanding process. Often the

Earning a Ph.D. is definately a demanding process. Often the

graduate school catalog suggests that candidates can recieve their

degrees after four years of work, but usally at least five years are

necessary.

The process consist of three stages. In the begining, the students do   5

two years of course work. Then they take several months studing for

the preliminary exam. This exam usually combines a written and an

oral test. Finally, the students begin writting their dissertations. A

dissertation presents an original opinion about a specialized topic of

intrest. The students aren't totaly finished until they defend their   10

dissertations in front of a panel of professors.

The Ph.D. candidates are suppose to do all of this while also trying

to make enough money to live on. Fellowships are not as generous

as they use to be; candidates must teach and work parttime while

preparing for their degrees. Nevertheless, the intellectual   15

excitement of the work, the satisfaction of acheiving the doctorate,

and the possibility of becoming a college professor drive many

students to complete the long process.

Capitalization

In the following sentences, add any necessary capitalization. For example, examine the following sentence:

my english professor, dr. fogarty, gave up the medical profession to become a college teacher.

This sentence should be corrected to read:

My English professor, Dr. Fogarty, gave up the medical profession to become a college teacher.

1. the inauguration of general washington as the first president took place in new york, the nation's original capital.

2. my friend albert is from deadsville, nevada, and doesn't know a chevrolet from a jeep.

3. our good neighbors, mr. and mrs. chaney, gave us their old set of encyclopaedia brittanica.

4. the first coca cola was bottled in vicksburg, mississippi, a small southern town, in 1894, by the biedenhorn family.

5. if you run a red light and cause a wreck, you may end up needing a doctor as well as a lawyer.

6. when marylynne graduated from high school last spring, her grandfather gave her his old cadillac.

7. on wednesday we saw a production of cat on a hot tin roof at the williamstown theater.

8. the reading list for professor schneider's american history 201 is two pages long.

9. last week we went to yankee stadium and saw the yankees beat the kansas city royals with a wild ninth-inning rally.

10. before caryl chessman was executed, he led a crusade to persuade the governor ai.d legislature of california to abolish capital punishment.

Abbreviations and Numbers

In the following paragraph, cross out any incorrect words and write corrections above them. For example, the first line should be corrected to read:

brothers
The Grimm ~~bros.~~ probably never expected that they would be best

The Grimm bros. probably never expected that they would be best

known for the 210 German fairy tales that they collected and

published in the yrs. between 1812 and 1850. Collecting fairy tales

was only one project undertaken by these 2 bros. They were

involved w/ all aspects of German cultural history. They collected   5

myths, legends, and medieval lit. They began an extensive dictionary

tracing the history of German words (similar to our *OED, the Oxford*

English Dictionary). Jacob Grimm discovered the links between the

Germanic languages of northern Europe (such as Eng.) & the

Romance languages of southern Europe (such as Spanish and   10

Italian). The Grimm bros. were also politically involved and were in a

group of 7 profs. who lost their jobs for disagreeing with the govt.

Writing Activity

Write directions for getting from school to your house. Write in terms of the mode of transportation that you normally take—for example, walking, driving, cycling, or taking the bus. Use numbers and abbreviations correctly. For example, you might write:

Corlears School is located at 324 West 15th Street next to El Cid Restaurant. When you leave the front door, turn right and walk to the corner of Eighth Avenue. . . .

Apostrophes

In the following sentences, add an apostrophe wherever necessary. For example, examine the following sentence:

No one sees her, but the narrators voice dominates the opening scene.

This sentence should be corrected to read:

No one sees her, but the narrator's voice dominates the opening scene.

1. The compositor says its not fair that he gets paid so little compared to the editor.

2. The new banks real estate holdings arent worth as much as the investors think.

3. In the summer, Larry spends his days exploring Mr. Ottomanellis meadow.

4. Depending on its half-life, a radioactive isotope can be dangerous for only a few seconds or for years.

5. A good nights sleep has its own quiet pleasures.

6. An article in todays newspaper reported that the president doesnt know of any improprieties on the part of the attorney general.

7. E.T. eats Reeses Pieces, not M&Ms.

8. "Odd, isnt it?" asks Fred. "Everyones dead and Ruths a

   millionaire."

9. When youve chopped all the ingredients, youre ready to start

   cooking Uncle Lens stir-fried shrimp.

10. Paulette doesnt put up with her managers rudeness.

Apostrophes

In the following paragraphs, add an apostrophe wherever necessary. For example, the first sentence should be corrected to read:

Ira's house is a world of its own.

Iras house is a world of its own. Downstairs lives his Uncle Abe. Ira

never enters Abes apartment. Upstairs, in the front room, Iras

parents sit for hours and talk. Pete, the old beagle, sprawls on a

hooked rug. In the back room, the boys play ping pong. Iras ping

pong table takes up the entire room, leaving a space of less than   5

eighteen inches for each of the players.

Beverly sits on her mothers porch and shells peas. The plunk, plunk,

plunk of the peas landing in the round tin tray blends in with the

raindrops drumming on the screens. Its Fourth of July and there will

be salmon and peas for breakfast. The rains untimely burst wont

affect her day. Shes sure that it will pass over in time for this   5

afternoons party and this evenings picnic.

Writing Activity

Write a paragraph about a house or apartment that you remember from childhood. Write in the present tense, as if you were there right now. Be sure to use some words that need apostrophes. Here is an example of how you might begin:

Every morning at Aunt Lillie's house on Farmer Street, Uncle Tabo cooks a big breakfast. We children can't wait to drink Uncle Tabo's big mugs of coffee milk. . . .

———————

Write a paragraph about a teacher—past or present—and his or her character, quirks, methods, clothes, etc. Be sure to use some words that need apostrophes. Here is an example of how you might begin:

I couldn't believe Mr. Small's love of math. He enjoyed writing examples on the board filled with x's and y's and impossible numbers. . . .

Consistent Pronouns

Correct each sentence to make the nouns, pronouns, and their verbs consistent. For example, look at the following sentence:

A great chef doesn't need to measure their ingredients.

There are several ways to correct this sentence:

A great chef doesn't need to measure the ingredients.

Great chefs don't need to measure their ingredients.

A great chef doesn't need to measure his or her ingredients.

1. If someone is disrupting the performance, they should be asked to leave.

2. A person should be careful not to break the yolk when you separate the egg.

3. To be a good waiter, one must be able to stay on their feet and smile for hours.

4. Even on cloudy days, when you may not seem to be exposed to sunlight, one should take precautions against ultraviolet rays.

5. When I wear unusual clothes, people look at you in a strange way.

6. A wise politician will be judicious with their promises.

7. When someone meets someone for the first time, they should try to remember his or her name.

8. I got mad; it does make you upset when people don't listen.

9. If an opera student is serious about their voice, they know the importance of breath control and they relax before a performance.

10. A person who is traveling should know not to schedule their vacation timetables too tightly.

Correct Pronouns

Correct the pronouns in the following sentences where necessary. In some cases, you may need to change the word order.

My mother and I
Example: ~~Me and my mother~~ would head out to school each day with me crying my eyes out.

1. He and his sister hope to practice law with my father and I.

2. For Amelia and I, both language majors, the year to study abroad was our chief consideration in choosing the college.

3. That controversy is for the Office of Consumers Affairs and her to resolve.

4. We sued her lawyer and her.

5. Oscar Hammerstein II may be our country's most influential lyricist; Richard Rodgers and him wrote America's most popular musical comedies.

6. Anna Karenina was torn between he and her husband.

7. Butch Cassidy and him robbed banks, from the American West to Bolivia.

8. For years, the Russians were our enemies, yet in our histories there are great similarities between them and us.

9. Between you and I, the jazz pianist should have won the competition.

10. The new insurance policy will require a smaller deductible for my husband and myself.

11. Frances and myself had the dirty job of cleaning all the shower curtains before she and I could pick up our paychecks.

12. Carl makes better cheesecake than me.

13. Carl works in the kitchen better with Margaret than with me.

14. Carl works in the kitchen better with Margaret than I do.

15. My sister is a better carpenter than me, but I play the guitar better than her.

16. Cleopatra loved Julius Caesar, but she loved Marc Antony even more than him.

17. Caesar conquered a great territory, but Napoleon conquered even

    more than him.

18. Guess whom is coming for dinner.

19. Whom should I say is calling?

20. We bought tickets for he and she for *The King and I*.

Writing Activity

Write a sentence using each of these pairs of pronouns correctly:

Example: Both Felicia and I like to swim in the lake at night.

1. Felicia and I

2. Mel and her

3. them and me

4. him and her

5. Lori and me

6. The committee and he

7. my friend and I

8. she and Benny

9. the players and him

10. my cousins and me

Vague Pronouns

Revise these sentences so that none of the pronouns is vague. For example, look at the following sentences:

Tim ordered Chinese food for us instead of pizza. This is my favorite take-out food.

There are two ways to interpret "this" in the sentence, and therefore two possible revisions:

Pizza is my favorite take-out food.
or
Chinese food is my favorite.

A better option is to combine sentences:

Tim ordered my favorite take-out food—Chinese—instead of pizza.

1. Use a mildly abrasive cleanser in which the tub enamel will be
   protected.

2. Food advertising rarely emphasizes vegetables, which is terrible for
   public health.

3. Ernie never eats raw peppers, mussels, or mustard. It makes him
   break out.

4. Over 75 percent of the inmates in New York State's prisons come
   from just seven neighborhoods in New York City, which has
   community leaders searching for solutions.

5. It is cheaper to buy two small plastic bottles of this detergent instead of one big one. From an environmental perspective, this makes no sense.

6. Mr. Charles chats with his employees on their first day of work, he helps them with their initial projects, and he supports their applications for promotion. This makes a big difference.

7. It was described to us in the story the route Peyton Farquhar would have taken, if in fact he did escape.

8. The evening light was dim despite the lantern on the porch, which made it hard to see.

9. The shoe gave Millard a callus on his foot that needed to be removed.

10. The sunlight reflected off the water and through the big pine tree which came right into the bedroom.

Recognizing Complete Sentences

In the following sentences, circle the subject and
underline the verb. For example, examine the following
sentence:

Many farm families in Normandy distill their own apple brandy.

This sentence should be marked as follows:

Many farm (families) in Normandy <u>distill</u> their own apple brandy.

Note that some sentences have more than one subject
and/or more than one verb.

1. Frank Lloyd Wright designed the furniture as well as the house.

2. Every single guest in the room stopped talking at the same time.

3. Every Friday morning, the bump-and-grind music shook the floor

   beneath our feet.

4. Joan Miro, known for painting in symbols, never painted on

   Sunday.

5. Arthur Penn directed Warren Beatty and Faye Dunaway in *Bonnie*

   and Clyde.

45

6. Every June, honeysuckle floods the roads of Virginia with perfume.

7. The twins placed an ad in the paper: "Last Day to Send in Your Dollar!"—and over a thousand people did.

8. I shut off the alarm clock and rolled onto my back.

9. There has never been a simple method for figuring federal income tax.

10. In the Battle of Gettysburg, over five thousand horses were killed.

11. Eddie Sauter, who is a major jazz arranger, recently did the arrangements for Stan Getz's album, *Focus*.

12. The cemetery flowers are either plastic or wilted.

13. That's the reason no one wants to lend her money anymore.

14. Lift the tab and pull back to remove the foil covering before inserting the cartridge into the printer cradle.

15. Introducing the dean, Marylin reversed his first and last names.

16. A number of mysteries incorporate recipes into their plots; Rex Stout's *Too Many Cooks* is one of the best.

17. Clean the upholstery regularly with a commercial leather cream, saddle soap, or a mild solution of vinegar and water.

18. In her paintings, Georgia O'Keefe gave sensuousness to flowers and bones.

19. Jane and Norman hired the catering service and engaged two bands for the wedding reception.

20. "Dumpster Diving" is the term for salvaging usable items from other people's trash; however, this practice is against the law because whatever is put out as garbage is the property of the Sanitation Department.

Recognizing Complete Sentences

WRITING ACTIVITY

Write two sentences in each of the following patterns. For example, you might write:

Simple Sentence:
The Queen Anne cherries were rotten.
There you go again.

Simple Sentence with introductory phrase:
In the morning, the clouds still hung low over the fields.
On our back porch stood the old refrigerator.

1. Simple sentence

2. Simple sentence with introductory phrase

3. Simple sentence with two subjects and one verb

4. Simple sentence with one subject and two verbs

5. Compound sentence connected with *and* or *but*

6. Compound sentence connected with *however*

7. Complex sentence using *when* or *after*

8. Complex sentence beginning with *If* or *Because*

9. Complex sentence using *if* or *because* in the middle

10. Compound-complex sentence

Sentence Fragments

In the exercises below, correct any fragments by attaching them to a sentence or by changing the wording. For example, examine the following:

> Madame Bovary abandons her husband and little daughter. In order to experience life to its fullest.

Here are some possible corrections:

> Madame Bovary abandons her husband and little daughter in order to experience life to its fullest.
> *or*
> Madame Bovary abandons her husband and little daughter. She wants to experience life to its fullest.
> *or*
> In order to experience life to its fullest, Madame Bovary abandons her husband and little daughter.

1. There are several subtypes of detective fiction. One being the hard-

   boiled school.

2. The prompter stands just backstage with the script. Whispering

   everybody's cues.

3. When he returned years later to help his mother move. He dug for

   the box exactly where he had buried it.

4. Swimming techniques vary depending on the choppiness of the

   water. Swimming in the ocean, for example, is different from other

   kinds of swimming.

5. The record sales figure for avocados each year is for the week of the Super Bowl. Presumably for all the guacamole dip—estimated at 6,000 tons.

6. The financial planner advised them to make a realistic budget. To figure out where their cash was going, then begin saving or investing in a mutual fund.

7. Gregor Mendel discovered the patterns of genetics. By observing how peas grow and reproduce.

8. In his later years, Matisse cut colored paper into ornate collages. Because he could no longer handle a paintbrush.

9. Einstein didn't clutter his mind with "useless" numbers. Such as his home telephone number.

10. I thought that Fred was dependable. That he would be at the restaurant at closing time to pick me up.

Sentence Fragments

The following paragraph contains sentence fragments. Correct them by changing either punctuation or wording. For example, the first line can be corrected to read:

The divorce rate was relatively low up until the 1950s.
or
Up until the 1950s, the divorce rate was relatively low.

The divorce rate was relatively low. Up until the 1950s. Then in the

1970s, it began to skyrocket. Many factors contributed to this:

People beginning to think more about individual fulfillment.

Religious restraints becoming looser. Women gaining a new

capacity to be economically independent. The divorce rate has   5

grown so much in the past two decades that each year for every two

marriages there is one divorce. Today it is not unusual for children

to see their parents divorce. Then see one or both of their parents

remarry and maybe even divorce again. Statistics say little about the

painful and frightening experience that children go through.   10

Especially the adolescents in the family. Many adjustments follow:

Living in a household with only one parent, a loss in economic and

social status, and sometimes a change of residence. Coping with a

divorce in the family is best done through counseling and support

groups. To assure that history will not necessarily repeat itself in the   15

future.

Run-on Sentences

Correct any run-on sentences by changing either punctuation or wording. For example, the first sentence can be corrected to read:

Tradition had it that the ghost appeared always on July 7th. It appeared only to someone who was not a member of the immediate family.

Note: You can also use a semicolon after *July 7th* instead of a period.

Tradition had it that the ghost appeared always on July 7th, it appeared only to someone who was not a member of the immediate family. On July 7, 1967, Dr. Rolph Dumaine was visiting the Beverleys, he was sitting on the back porch. Ann Beverley served him his second mint julep; she began to describe the death of her   5 favorite dog Jeb Stuart. Suddenly Dr. Dumaine turned to her and asked, in a voice mildly curious, "Who is the woman with the long red hair, I haven't been introduced to her?" Ann Beverley's face lost all its expression. It was as blank as a March day's sky. "Red-haired, you say? What was she wearing?" "A yellow dress with a wide green   10 belt, she was heading toward the library." Ann Beverley fainted, the old dog began to howl. Dr. Dumaine felt a cold sweat trickle down his neck. "It's July 7th," he said to himself. "Why do I feel cold?"

Run-on Sentences

Correct any run-on sentences in the exercises below by adding punctuation or by changing the wording where necessary. For example, look at the following sentence:

Mozart deserved a better end, he died penniless and was buried in an unmarked grave.

Here are three possible corrections:

Mozart deserved a better end. He died penniless and was buried in an unmarked grave.
or
Mozart deserved a better end; however, he died penniless and was buried in an unmarked grave.
or
Mozart, who deserved a better end, died penniless and was buried in an unmarked grave.

1. The male emperor penguin holds the egg on top of his feet, he does this for two months at a time.

2. There is a secret to cooking good hard-boiled eggs, they should never be boiled.

3. A kitten can sleep in one spot for hours, other times it prowls, stalks, pounces on imaginary prey.

4. An inexpensive styrofoam cooler saved their lives, they clung to it for hours after their boat capsized.

53

5. You're supposed to blanch vegetables, you're not supposed to bleach them.

6. Pigeons manage to survive in the city they somehow make it through the coldest winters.

7. The loggerhead turtle scraped heavily up to the dunes, her flippers left a distinctive cross-hatching in the wet sand.

8. Consumer groups want no pesticides, however extension agents seek reduced use of pesticides, just enough to protect food from disease.

9. In the highest mountains of Peru, steam engines actually outperformed diesel engines, evidently, they required less oxygen.

10. When he was twenty-two, Charles Darwin set sail on *The Beagle*, he didn't return to England for five years.

Sentence Fragments and Run-on Sentences

Correct the punctuation of these sentences as necessary, without changing the wording. For example, look at the following:

Because children are natural artists. They should be encouraged to draw and paint.

The punctuation and capitalization can be corrected to read:

Because children are natural artists, they should be encouraged to draw and paint.

1. Most children are natural artists however they often quit drawing when they grow up.

2. Although most children are natural artists they often quit drawing when they grow up.

3. Most children are natural artists although they often quit drawing when they grow up.

4. Children should be encouraged to draw and paint because most of them are natural artists.

5. Most children are natural artists but they often lose this talent. As they grow up.

6. Most children are natural artists therefore they should be encouraged to draw and paint.

7. Most children are natural artists but they often quit drawing when they grow up.

8. Since most children are natural artists they should be encouraged to draw and paint.

9. Most children are natural artists they should be encouraged to draw and paint.

10. Children should be encouraged to draw and paint. Since most of them are natural artists.

Sentence Fragments and Run-on Sentences

Continue these sentences and then add any necessary punctuation. You may need to change capitalization, but do not change the original wording. For example, if the exercise sentence reads

I went to the zoo although

you might write

I went to the zoo although I had only an hour.
or
I went to the zoo. Although I had only an hour, I saw all the birds.

1. I went to the zoo however

2. I went to the zoo but

3. I went to the zoo it's

4. I went to the zoo then

5. I went to the zoo especially

6. I try to avoid fatty foods therefore

7. I try to avoid fatty foods which

8. I try to avoid fatty foods nevertheless

9. I try to avoid fatty foods such as

10. I try to avoid fatty foods if

Sentence Fragments and Run-on Sentences

Correct any sentence fragments or run-on sentences by changing punctuation or wording. For example, the first sentence can be corrected several ways:

Some people can never pass a pinball machine without stopping to play a game. They get hooked on the challenge of keeping the ball on the move.

(Note: You can also use a semicolon after "game" instead of a period.)

or

Some people can never pass a pinball machine without stopping to play a game because they get hooked on the challenge of keeping the ball on the move.

Some people can never pass a pinball machine without stopping to

play a game they get hooked on the challenge of keeping the ball on

the move. And racking up a score of a million points or more. The

builders of these machines are smart; they add pops and whirrs,

blinking lights, music, and sometimes even an alluring voice. Which   5

calls out to you to come and play a game. If you give in to the

temptation and insert a quarter into the slot. A round metallic ball

will come up on the right side of the machine. Then the game starts.

You pull a plunger-like instrument back. This sets the ball in motion.

It scoots around the surface doing many things   10 at one time. Like hitting bumpers, sounding off bells and chimes, turning on lights. Eventually the ball will roll to the flippers, which you control with your fingers. Your skill in using the flippers will often determine whether or not you will tally up a high score. A game of pinball can be mesmerizing if you're not careful, you might find that   15 an entire day has passed and you're still at it, trying to outdo your last score.

Sentence Fragments and Run-on Sentences

In the paragraph below, correct any sentence fragments or run-on sentences by changing punctuation or wording. For example, the first two lines can be corrected several ways:

Fashion photography, once considered a commercial endeavor, has now become an important art form, which is exhibited in museums and galleries.
or
Fashion photography, once considered a commercial endeavor, has now become an important art form. It is exhibited in museums and galleries.
or
Once considered a commercial endeavor, fashion photography has now become an important art form. It is exhibited in museums and galleries.

Fashion photography, once considered a commercial endeavor, has

now become an important art form. Which is exhibited in museums

and galleries. The field of fashion photography has attracted some

of the most outstanding photographers of the twentieth century.

Like Richard Avedon, for example. Avedon is a brilliant photographer   5

in many areas, but he has made his biggest mark in the field of

fashion photography in fact, Avedon has been one of the driving

forces in turning fashion photography into a respected art form. His

work has virtually dominated the field since the late 1950s, and it is

still published in almost every fashion   10

magazine around the world. His photographs are bold and

straightforward, he projects a cool, clean, thoroughly modern image.

His style has changed with the times over the years. That is why he

is still considered one of the best fashion photographers of the day.

And a legend in his own right.   15

Writing Activities

————————

Write a description of a place that you love to visit. Tell what you love about being there. Check your description carefully for sentence fragments and run-on sentences. Here is an example of how you might begin:

> In the fall I hike down the Natchez Trace to lie on the haystacks that farmers leave in the fields. Lounging up on top, reading a book, drinking Coca-Cola, and muching parched peanuts, I feel like I'm in heaven. . . .

————————

Write a paragraph telling about the most dramatic weather you've ever been in. Use each of these words or phrases at least once:

however   it   for example   therefore   they

Then go back and check your punctuation carefully for sentence fragments and run-on sentences. Here is an example of how you might begin:

> As soon as we heard the flood warnings, we left the square dance; however, we didn't leave soon enough. It was raining so hard that the windshield washers couldn't keep pace, so Pat had to lean out the car window to see. . . .

Commas

Insert commas where necessary. For example, look at the following sentence:

Palermo Sicily once a great world city is now a remote little-used seaport.

This sentence should be corrected to read:

Palermo, Sicily, once a great world city, is now a remote, little-used seaport.

1. Allied troops landed in Normandy on June 6 1944 and began the

   liberation of Nazi-occupied France.

2. Josef Breuer Sigmund Freud's teacher developed the "talking cure"

   in which patients were encouraged to tell their life stories.

3. The Center for Astrological Studies was established in 1972 in

   Lancaster Vermont two hundred miles north of New York.

4. We explored Ferry Beach and waded in the pools between sand

   bars.

5. Mix in the chocolate chips nuts and raisins and refrigerate the

   dough for three hours before baking.

6. The new technician put the blank tape into the machine and pushed several buttons hoping to record the entire concert.

7. When we left the game we drove for twelve hours straight in order to get home by the next day.

8. Even though high-bush blueberries are easier to pick low- bush blueberries taste better.

9. James Agee a writer and Walker Evans a photographer published a book entitled *Let Us Now Praise Famous Men* about tenant farm families during the depression.

10. In the movie *Wall Street* Darryl Hannah and Michael Douglas soften the ice cream in the microwave.

Commas

In the sentences below, take the word or words from the second line and insert them into the first line. Add commas where necessary. For example, look at the following:

Fashion photography has now become an art form.
once considered a commercial endeavor

The new sentence should read:

Fashion photography, once considered a commercial endeavor, has now become an art form.

1. Edward Hopper specialized in both rural and urban settings.

   the American painter

2. Adam Smith argued in favor of a free market.

   who wrote *The Wealth of Nations* in 1776

3. Electrotherapy enables athletes to increase muscle size

   without the use of steroids.

   however

4. Robert Frost was born in San Francisco and lived there until he

   was ten years old.

   California

5. No plan offered a yield of more than 7 percent.

   to my knowledge

6. Aunt Margaret had strange feelings when she entered the Vatican.

   who always claimed to be an atheist

7. Passive smoke can be a severe threat to health.

   which is inhaled while in the company of smokers

8. Studies have shown that you shouldn't lie down after eating.

   if you want to lose weight

9. Retailers are depending on the consumers to spend more freely.

   always hopeful that sales will increase during the

   Christmas holidays

10. The report was vigorously protested by the committee.

    a seventy-two-page attack on the administration

Commas

For each of the following phrases, write a sentence and insert the phrase into the middle. Then check that you have added the necessary commas.

Example: as it turned out
The rally, as it turned out, was postponed until thefollowing week.

1. as it turned out

2. however

3. my best friend

4. who won the contest

5. July 4 1989

6. a wrestler

7. Boise Idaho

8. which used to be purple

9. on the other hand

10. in my opinion

Writing Activities

In a paragraph, write a recipe, telling how to prepare one of your favorite dishes. Then check the punctuation of your paragraph. For example, you might begin:

To make nachos grandes, first take three ten-inch flour tortillas and cover them with grated Monterey Jack cheese. Next, chop one medium onion, three plum tomatoes, and one jalapeño pepper. . . .

Write a brief autobiography, telling where and when you were born, who your immediate family members are, and the names of some of the places where you have lived or visited. Then check punctuation.

Semicolons and Colons

In the following sentences, insert semicolons or colons. You may also need to insert commas where appropriate. For example, look at the following sentence:

> Wood-burning stoves have at least one ash lip it is the little shelf that catches the ashes when the door is opened to add more wood.

This sentence can be corrected to read:

> Wood-burning stoves have at least one ash lip; it is the little shelf that catches the ashes when the door is opened to add more wood.

1. Students should take off from college for a year for three reasons to learn the demands of the work world to consider the field they want to study and to get a break after years of constant schooling.

2. If two sides of a triangle are equal the angles opposite those sides will be equal.

3. When Pooh arrived at Piglet's house he expected to find him but Piglet had just left for Pooh's house.

4. The Adirondacks which until the late 1800s were still remote and undeveloped became the unlikely playground of the very rich.

5. For example a lawyer begins at a salary of $65000 in one top firm.

6. Michael was thrilled with his Gore-Tex rain gear especially when he could bicycle home in the rain.

7. Steel magnate Henry Frick ruthlessly opposed his workers' efforts to organize however the same man was one of the world's great art collectors.

8. The man standing in front of me in the supermarket line complained so much about the long line that he annoyed everyone.

9. This school district has very flexible personnel policies for example family leave is available to anyone who must care for a relative.

10. Thoreau worked at a variety of jobs one being handyman for the Emersons.

Semicolons and Colons

In the following sentences, insert semicolons or colons. You may also need to insert commas where appropriate. For example, examine the following sentence:

They placed an order for the following items three bags of fertilizer two garden rakes and four trays each of tomato green pepper and cucumber plants.

This sentence should be corrected to read:

They placed an order for the following items: three bags of fertilizer; two garden rakes; and four trays each of tomato, green pepper, and cucumber plants.

1. Agatha went to the supermarket and spent one hundred dollars on

   food even though she knew that they had enough at home.

2. Agatha knew that they already had enough food nevertheless she

   went to the supermarket and spent one hundred dollars.

3. Agatha bought all sorts of things she didn't need papaya juice

   sardines and maraschino cherries.

4. She bought some very expensive items such as caviar olive paste

   and saffron.

71

5. Half of the items in her shopping cart were blue blueberries blue cheese and blue corn chips not to mention blue plastic forks blue paper napkins blue soap and blue toilet paper.

6. Notable guests at the reception included Herman Bernhof the furrier Maisie Fason the ballet teacher Nate Maybloom the second baseman and Marietta Daley the mayor.

7. Most of the guests were trying to impress the mayor however Nate Maybloom ignored her.

8. The newlyweds inherited furniture from their rich Aunt Sophie therefore they never got to choose their own style.

9. The newlyweds inherited furniture from their rich Aunt Sophie so they never got to choose their own style.

10. Their tiny apartment was crammed with large pieces of furniture for example they had three matching gray armchairs.

Dashes and Parentheses

In the following sentences, add dashes where necessary. For example, examine the following sentence:

Rushing into the shop he's always in a hurry he ran smack into a mannequin.

This sentence should be corrected to read:

Rushing into the shop—he's always in a hurry—he ran smack into a mannequin.

1. He was so taken by *Pride and Prejudice* that he spent the summer reading all of Jane Austen's novels every single one.

2. Archie marched into the room and are you ready for this? sprayed the marble statue pink.

3. There are times when I want to be totally alone times when I need to hear my own thoughts.

4. The record snowfall over nine feet in some areas will do much to alleviate the drought.

5. My favorite songs are all by Steely Dan "Turn That Heartbeat Over Again," "Only a Fool Would Say That," and "Dr. Wu."

In the following sentences, add parentheses and any other necessary punctuation. For example, examine the following sentence:

Snakes nonpoisonous ones are my favorite pets.

This sentence should be corrected to read:

Snakes (nonpoisonous ones) are my favorite pets.

1. I spoke with Victor a pet shop owner who had advertised a boa

   constrictor in the newspaper.

2. I was eager to see the snake snakes have fascinated me since

   childhood

3. Victor brought the snake at six o'clock the store closes at five so

   the snake could see her new habitat.

4. I offered both of them dinner not the same menu, of course

5. I knew I'd found my pet when can you believe this I saw the snake

   smile with satisfaction.

Quotation Marks

This exercise is based on the following quotation—the opening paragraph of *Invisible Man*, by Ralph Ellison:

> I am an invisible man. No, I am not a spook like those who haunted Edgar Allan Poe; nor am I one of your Hollywood movie ectoplasms. I am a man of substance, of flesh and bone, fiber and liquids—and I might even be said to possess a mind. I am invisible, understand, simply because people refuse to see me. Like the bodiless heads you see sometimes in circus sideshows, it is as though I have been surrounded by mirrors of hard, distorting glass. When they approach me they see only my surroundings, themselves, or figments of their imagination—indeed, everything and anything except me.

Add punctuation marks where necessary in the following sentences. For example, look at the following sentence:

Ralph Ellison begins his novel with a surprising and memorable line I am an invisible man

This sentence should be corrected to read:

Ralph Ellison begins his novel with a surprising and memorable line: "I am an invisible man."

1. Ellison writes I am invisible, understand, simply because people

   refuse to see me

2. When the narrator says that he is not one of your Hollywood- movie

   ectoplasms the reader knows that the narrator is not a science

   fiction creature

3. Invisible in this context means not really being seen as a person

4. Ellison uses the word invisible to describe the feelings of African Americans in certain situations

5. When Ellison says that he is an invisible man he speaks not only for African Americans but for anyone in our society who feels powerless and anonymous

6. Ellison says what many of us at times want to assert I am a man of substance, of flesh and bone, fiber and liquids—and I might even be said to possess a mind

7. The narrator explains right from the start that he is not a spook like those who haunted Edgar Allan Poe

8. Why did Ellison choose the word invisible?

9. Ellison explains that when they approach me, they see everything and anything except me

10. Ellison explains I am not a spook like those who haunted Edgar

    Allan Poe the author of many horror stories; nor am I one of your

    Hollywood-movie ectoplasms.

Writing Activity

Write down a conversation between you and one other person in which you are trying to make a decision. Make sure that you start a new paragraph every time a different person speaks.

Here is an example of how you might begin:

> "I don't know. I don't feel like having Chinese food tonight." I stood on the corner of 18th Street and looked at Spring Joy's neon sign.
> "Well, MaryAnn's is down the block," Jay said. "Does Mexican sound good?" . . .

Underlining or Quoting Titles

In the following sentences, add underlining or quotation marks and any other necessary punctuation.

For example, look at the following sentence:

I recently read an article in Reminisce magazine called Hobo Symbols.

This sentence should be corrected to read:

I recently read an article in <u>Reminisce</u> magazine called "Hobo Symbols."

1. One of the most chilling of all the antiwar novels is Dalton

   Trumbo's Johnny Got His Gun.

2. The short story A Conversation with My Father by Grace Paley can

   be found in Enormous Changes at the Last Minute.

3. Many of the great old magazines such as Collier's, Saturday

   Evening Post, and Scribner's have disappeared.

4. Loretta and Tim went to see Alfred Hitchcock's The Birds at the

   revival movie theater.

5. John Keats's best-known poem is probably Ode on a Grecian Urn.

6. Many people don't known that The Doors wrote the pop song Light My Fire.

7. Last Monday's Evening Herald carried a controversial article entitled Charitable Organizations Give Away Very Little Money.

8. Ralph, a ten-year-old, and Molly, an eight-year-old, are the main characters in the novel The Mountain Lion by Jean Stafford.

9. The old I Love Lucy television shows are still as funny as they were when they were filmed in the 1950s.

10. A recent article Beneath the Sea, Beyond the Ages in the flight magazine American Way describes how archaeologists discovered a 3,000-year-old ship at the bottom of the Mediterranean Sea.

Writing Activity

Write a newspaper article about a make-believe awards ceremony for writers. Make up authors' names and titles for awards in the following categories. Be sure to use quotation marks and underlining as needed.

Here is a way you might begin:

> The surprising winner at last night's literary banquet sponsored by the <u>Rawlings Times</u> was Caroline Greer for her narrative poem "A Day on Golden Street."

Best novel

Best short story

Best film script

Best magazine article

Best song

Verb Tenses

In the following sentences, write the correct form of the verb in the space provided. For example, examine the following sentence:

The speaker _____ the crowd to its feet at last week's rally.
           bring

This sentence should be corrected to read:

The speaker __*brought*__ the crowd to its feet at last week's rally.
           bring

1. More books have been _____ about the Civil War
                   write

   than about any other subject.

2. The art exhibit is poorly _____.
                   arrange

3. Before it _____, the kiln temperature _____
          explode   exceed

   the federal safety limit.

4. The mockingbirds in Brooklyn are now _____
                        mimic

   car alarms.

5. He had only half _____ the apple when he
               eat

   _____ the worm.
    see

6. Uncle Lute _____ sick even though he didn't
   get

   _____ the local water.
   drink

7. Akira Kurosawa _____ to be a painter before he
   use

   _____ a famous film director.
   become

8. If disposal expenses were _____, this flashlight
   include

   would _____ more than its competitors.
   cost

9. As his ship was _____ down, John Paul Jones
   go

   declared, "I have not yet _____ to fight!"
   begin

10. I would _____ to have something decadent and
    love

   chocolate because I haven't _____ anything sweet in
   have

   three weeks.

Writing Activity

Write a paragraph about a plan you have. Pay special attention to each verb form. For example, you might begin:

Next week, I will be driving up to Oregon. I plan to take two days. . . .

———————

Write a paragraph about something that has been going on recently in your life. For example, you might begin:

Since February, I have been dieting and exercising. I began by drinking a glass of water before every meal and by walking two miles a day.

———————

Write a paragraph about something that happened when you were in elementary school. For example, you might begin:

The event I remember most vividly from the second grade is the play we put on. Mrs. Tucker chose me to play the Spanish dancer.

Shifting Verb Tenses

In the following sentences, cross out any incorrect verb and write the correction above it.

Example: Employers look for the most competent applicant they

can
~~could~~ find.

1. The project failed but they done the best they could with the

   amount of time they had.

2. When the end of the war was announced, the people would dance

   in the streets.

3. Flaubert takes us right up to the moment when Rodolphe had

   kissed Emma and then writes, ". . ."

4. At the beginning of *The Nutcracker*, Uncle Drosselmeyer brings

   Maria a toy nutcracker; and when everyone has gone to bed, it

   came to life.

5. Before he hitchhiked home, Todd and his mother had already went

   to six national parks.

6. The Russian ice dancers had just earned a perfect score when the French couple had taken the ice.

7. When Hamlet learns that his uncle has murdered his father, at first he couldn't find a way to take action.

8. In the fossil of the cat there was a section that has been repaired.

9. Ever since I could remember, I was taught to work hard and try to excel at what I could do.

10. If Thurgood Marshall hadn't played hooky, he wouldn't have had to memorize the Constitution as punishment, and then he might not have become a great Supreme Court justice.

Shifting Verb Tenses

In the following story, cross out any incorrect verb and write in the correction above it.

For example, the first two sentences should be corrected to read:

When I went away to finish my degree, three students rented my

came
house. And when I ~~had come~~ back, they had four cats.

When I went away to finish my degree, three students rented my

house. And when I had come back, they had four cats. Two of them,

Narnia and Mary, were pregnant; Grendel already delivered a litter.

The fourth, called "Kitten," was the only cat they had kept from that

litter.

I demanded that my tenants had found a solution to this cat

population explosion, and one by one, the cats did find homes.

Grendel went to a farm as a barn cat because she was a skillful

hunter. Narnia and Mary each joined families with children. Only

Kitten had been left. Just three months old, she was the cutest and

the liveliest, but the boyfriend of one of the students had wanted

her.

Although I had always scorned people's devotion to their pets,

I secretly had wanted to keep Kitten. I kept wishing that the

boyfriend would let me have her. He didn't want to, but he couldn't

find an apartment that accepted pets, and his parents wouldn't let

him bring her home; so I inherited Kitten and had named her Kasha.

She has lived with me to this day.

Writing Activities

Write a story about a holiday. Use as many of the
following words as you can, not necessarily in this
order:

ate	did	went	saw	wrote
eaten	done	gone	seen	written

For example, you might begin:

I have never written about Passover at Uncle Murray and Aunt
Agnes's house. When I was small, we went there every spring for the
holiday . . .

Write a paragraph about a photograph of you or
members of your family. Tell the story behind it
and what was happening in it. Then go back and
check your verb tenses. For example, you might
begin:

Our family reunion photograph was taken in front of a weeping willow
tree. I remember that my cousin Evelyn had come to call us for the
photographs. . . .

Verb Agreement

In the following sentences, correct any verb agreement errors.

comes
Example: One of the guitarists always ~~come~~ in late.

1. The drummer and the saxophonist has placed in the final music competition.

2. Either the drummer or the saxophonist has placed in the final music competition.

3. Either the drummers or the saxophonists has placed in the final music competition.

4. The check, which is in the mail, covers my last payment.

5. The check that was listed in today's receipts cover my last payment.

6. The pattern of the rugs in the lobby are typical of Bokhara.

7. The patterns in the rug are typical of Bokhara.

8. Mixing over twenty different soundtracks are not unusual in

   commercial films.

9. At the conference, there was a moderator and three panelists who

   spoke in favor of the proposal.

10. Everyone in these neighborhoods speak Italian.

Verb Agreement

In the following sentences, correct any verb agreement errors.

demonstrates
Example: Each of the paintings ~~demonstrate~~ a different use of shadow.

1. On the new VCRs, programming several shows is now much easier than on older models.

2. Counting words in your papers take too much time.

3. One reason for reducing the use of pesticides is to cut costs.

4. Each of the rolls of film was labeled on the envelope.

5. Hemingway's use of simple words and sentences continue to influence many writers.

6. There was a physician and two research scientists on the commission.

7. Bryant's reasons for advocating the gold standard was partly religious and partly economic.

8. The two points made by the prosecutor was very weak in the view of the jurors.

9. The first of the guests always seem to arrive twenty minutes early.

10. The members of the board of trustees and the staff of the Arts Council supports the annual fund-raiser enthusiastically.

Writing Activity

Write a sentence beginning with each of these words or phrases. As a subject, write about a performance, such as a concert or play, using the present tense. Then check your sentences for verb agreement. For example, you might write:

Everybody is usually impatient for the show to begin.

1. Everybody

2. Neither my friend nor I

3. Each of us

4. Anyone

5. All of us

6. Either

7. There are

8. One of the performers

9. The lights and the sound system

10. Clapping my hands

Word Endings

Check each word in every sentence and change any incorrect endings. For example, look at the following sentence:

This movie director like to scare people with special effect.

This sentence can be corrected to read:

This movie director likes to scare people with special effects.

1. At his inauguration, President William Henry Harrison caught a

    cold, which turned into pneumonia, and he die a month later.

2. Helen Keller was so determine to learn that she managed to live a

    very full life with seemingly impossible handicap.

3. He didn't save the World Arts Organization any money, but he did

    computerized their accounts.

4. I wish I would have know that you were both going to the Sound

    and Light Show.

5. Some people like to work out in health club because it relieve their

   tension after a hard day's work.

6. Enormous sum of money could have been save if only the dangers

   of asbestos had been known.

7. When my aunt arrived home, she was shock to see what had

   happened.

8. Joy is very prejudice against various group; her prejudice makes

   her very unpleasant to be around.

9. Analysis shows that this brand of apple muffin contain only two

   grams of apple.

10. As the model walks into the room, she twirl around, lift her skirt,

    and shows polka-dot stockings which end above the knee.

Word Endings

-ED

Read the following paragraphs and correct any errors in word endings by adding or deleting *-d* or *-ed*. In some cases you may need to change the spelling of the verb. For example, the first sentence should be corrected to read:

qualified
Last summer, Glenda ~~qualify~~ to serve as a grand juror.

Last summer, Glenda qualify to serve as a grand juror. She and

twenty-two others had to decided whether suspects should be tried.

Usually, the prosecutor presented two witnesses. The first, the

victim of the crime, would tell what happen. This witness could tell

only what he or she actually saw and heard. In many cases, the

victim actually chased the suspect while calling for help. The second

witness was usually the police officer who made the arrest. Both

witnesses would testified that the victim saw the suspect in the

officer's custody. Normally, this evidence was enough to justify a

trial.

But sometimes the suspect himself or herself did testify. The

suspect usually claimed to have just been passing by when the police suddenly make the arrest. These cases were harder to

decide. The jurors didn't want to bring an innocent person to trial.

Usually, however, the suspect blundered into contradicting his or her

testimony or lying in an obvious way. When the story didn't make

sense, or when the suspect had been seen by two eye- witnesses,

the jurors had less of a problem. After much discussion, these

suspects, too, were usually sent to trial.

Word Endings

-s

Read the following paragraph and correct any errors in word endings by adding or deleting -*s*, -*es*, or *'s*. For example, the first sentence should be corrected to read:

pleasures
One of Marv's small ~~pleasure~~ is to go out to breakfast.

One of Marv's small pleasure is to go out to breakfast. His favorite

places are coffee shops and truck stop. He find that the cheaper the

breakfast, the better the food usually is. He always order the same

breakfast: two eggs over easy, toast, orange juice, home fries, and

coffee. After he goes to the same place a few time, he gets to knows

the person who waits on him, whose style is part of Marv pleasant

experience. Sometimes he might get into a brief conversations with

the waiter; other times he simply will read his newspaper while he

sip his coffee. He enjoys not only the food and the person behind

the counter, but also the feeling of independence. He likes to do

thing on his own and to observes the world. At these times, he can

feel that he is a man—free and friendly, with a few dollar in his

pocket and a little time for himself.

Writing Activities

Choose any short sample of writing—for example, from a newspaper, magazine, or book—and read it aloud slowly, stressing the end of each word. Then copy the passage word for word and check for accuracy. (To be effective, this exercise should be done regularly.)

Write a page about an occurrence that you observed recently in a public place and tell what you thought about it. Then check every noun for *–s* and every verb for *–ed, -s, or –ing.* For example, you might begin:

> I was waiting in line to buy tickets for the concert. I watched a tall man carrying several packages approaching me. . . .

Write down the story of a television show that you watched recently. Include words ending in *–s, -ed,* and *–ing.* For example, you might begin:

> In the beginning, several police detectives in plain clothes were gathered at the police station drinking coffee and joking. . . .

Tangled Sentences

PARALLEL STRUCTURE

Add or delete words to make the parts of the sentence parallel. For example, look at the following sentence:

> Members of the committee began every meeting by catching up on each other's news, drinking coffee, and they agreed on an agenda.

This sentence can be corrected to read:

> Members of the committee began every meeting by catching up on each other's news, drinking coffee, and agreeing on an agenda.

1. Lord Byron's travels took him to France, Switzerland, Italy, and to Greece, where he died.

2. The horse galloped, jumped, and a fallen log tripped him up.

3. In addition to watering and fertilizing your plants, try talking to them and to play classical music.

4. The plumber botched the job badly, demanded to be paid, and was asking to use the phone.

5. To keep vegetables fresh, first rinse them in fresh water, then soak them in water with a few drops of Clorox, and finally one should rinse them again.

6. He appreciates the force of her arguments, the subtlety of her wit, and she obviously had a grasp of the facts.

7. To enhance visual communications on campus, colors used on signs need to be consistent, their styles should be uniform, and use the same materials for the signs.

8. In spite of her beauty, in spite of her impeccable manners and her grace, she didn't get what she wanted.

9. We will not put up with the situation any longer, will not pretend it doesn't exist, and not be treated in such a manner.

10. The moon can be observed by looking into a telescope or the naked eye.

Tangled Sentences

DANGLERS

Revise these sentences to correct the danglers. For example, look at the following sentence:

Smelling powerfully of fish, I walked past the seafood market.

This sentence can be corrected to read:

I walked past the seafood market, which smelled powerfully of fish.

1. Last seen wandering vaguely, the police couldn't find Alex's uncle.

2. Rolled into the operating room on a stretcher, the doctor prepared to remove my appendix.

3. Pat was thrilled to finally get a bite but, reeling up the line too fast, the trout got away.

4. A colossal catastrophe, the firemen arrived at the burning chemistry lab.

5. Hanging from the chandelier, the astonished diners looked up at Mark's boots.

6. Walking across the soft spring soil, worm holes dimpled the surface.

7. Glowering at each other from opposite corners, the crowd grew tense waiting for the two wrestlers to meet in the middle of the ring.

8. A famous actor, the audience was nevertheless bored by Edward Kean's performance that night.

9. A meticulous planner, the suspense films of Alfred Hitchcock were designed so thoroughly that he did not even need to see the finished film to know what it looked like.

10. To practice medicine, not only a medical degree but internship and residency are required.

Tangled Sentences

MIXED SENTENCE PATTERNS

Revise these sentences to correct the mixed sentence patterns. For example, look at the following sentence:

In history has shown that people who speak out can change the world.

Here are two ways that this sentence can be revised:

History has shown that people who speak out can change the world.
or
In history we learn that people who speak out can change the world.

1. If one of the train engineers has been taking drugs could cause a fatal crash.

2. By taking a brisk two-mile walk every day will help you live to be a hundred.

3. As for an example would be Nick Nolte, who has played a wide range of roles.

4. To assemble this container using ordinary household glue.

5. In New York, one of the world's most populated cities, has a broad sample of the human species.

6. For those students who are having trouble keeping up they will

   receive special assistance.

7. Although many people may think I'm strange, but one of the things

   I love to do most is go to work.

8. By having ample money, it gives you confidence to do things on the

   spur of the moment.

9. In this article says how acid rain is affecting our country's forests.

10. Because of my aunt's generosity allowed me to attend the college

    of my choice.

For Extra Measure

CUMULATIVE EXERCISE ON CORRECTNESS

The following story contains errors. Make the necessary changes in spelling, verb or pronoun agreement, sentence correctness, and punctuation.

Every summer Dorothy use to go to the beach in South Carolina, she

rented the bottom floor of a house. One night, she walked down to

the dunes. She sat and for a long time watched the waves. Just

when she was already to go back to the house, she noticed a black

thing by the water. At first she thought it was a dog, however it was   5

to big. Besides, she hadn't seen it come from anywhere, it just

appeared. Then it started up the beach. Right towards her. Now she

saw what it was—a giant sea turtle. It marched up to the dune right

below her and began to dig. Dorothy realized that this was a female

about to lay her eggs. For over an hour the turtle used her flippers to   10

dig. Then to cover the eggs with alot of sand. Finally, she started

back to the sea. Dorothy walked along beside her, she even touched

the shell with her hand. It felt like stone. The turtle was obviously

exhausted. She moved very slowly. Stopping for long pauses. When

she reached the sea however she came to life. With one swift   15

motion she disappeared into the waves.

CUMULATIVE EXERCISE ON CORRECTNESS

The following paragraphs contain errors. Make the necessary changes in spelling, verb or pronoun agreement, sentence correctness, and punctuation.

Antarctica is one of the most desolate places on earth, it's impossibly cold climate makes it uninhabitable. Scientist go their to study many animals. Such as the penguins, seabirds, whales, and seals. Scientists also study the icebergs, the atmosphere, and observe the stars. They even drill down into the icebergs and   5 remove long samples of ice, called ice cores so that they can study the air deep in the ice, signs of volcanic eruptions, and pollen grains that are thousands of years old. They also have found that fish are able to live way below sea level, down under the icebergs.

But noone lives there for to long. Everything, all food and supplies,   10 have to be flown in. The weather is brutal. Temperatures can drop to minus one hundred degrees Fahrenheit and winds can reach two hundred miles per hour. When the sun sinks below the horizon in the middle of March Antarctica is left in the dark for six mos. The only light comes from the moon, the stars and the southern lights   15 (Aurora Australis).

CUMULATIVE EXERCISE ON CORRECTNESS

The following paragraph contains errors. Make the necessary changes in spelling, verb or pronoun agreement, sentence correctness, and punctuation.

Every since celebrities such as Elvis Presley, Marlon Brando, and

James Dean started riding motorcycles motorcycles have gained

alot of popularity amoung everyday people. But motorcycles got a

bad name back when the Hell's Angels formed in the nineteen

sixties. Since then, motorcycle riders had been stereotyped as black-   5

leather criminals. People driving cars often look with contempt at

motorcycle riders. Wondering how any one could ride a death

machine and endanger the lives of others as well as themselves.

Sure some crimes have been commited by people on motorcycles;

but probably many more crimes have been committed by people in   10

automobiles. And few people know about the many motorcycle

groups which makes it a point to do good deeds. For example every

December many groups get together for a toy run: They buy toys for

children in hospitals. And deliver them Christmas Eve. Many riders

have individually rescued people being mugged.   15

Unfortunatly, these good deeds are rarely reported on the front page

of newspapers. So the general public continues to stereotype people

on motorcycles. Yet the majority of motorcycle ridders are just

regular people who like the feeling of freedom they get. Zipping

down the highway with the wind in their faces.   [20]

CUMULATIVE EXERCISE ON CORRECTNESS

The following paragraphs contain errors. Make the necessary changes in spelling, verb or pronoun agreement, and punctuation.

When many people hear the word *Oz*, they usually think only of the

movie, filmed in 1939 staring Judy Garland. What they dont know is

that there are fourteen books that take place in the magical

kingdom of Oz. All of them written by L. Frank Baum. After the

success of *The Wonderful Wizard of Oz,* published in 1900, Mr.   5

Baum received many pleas from children asking him to write more

about Dorothys adventures with her friends from Oz; the Scarecrow,

the Tin Man, the Cowardly Lion, and the Wizard.

In 1904, four years after he published the first Oz book Mr. Baum

wrote an author's note with his second book, The Marvelous Land of   10

Oz. In it he says he thought he was all finished with Oz, but to many

children wrote letters to him and one little girl—who's name was

Dorothy—came a long way to visit him and had asked him in person

to write more about Dorothy. He told her that if he recieved one

thousand letters from little girls he would write another book. He   15

received many thousands and ended up writing fourteen Oz

books, often using ideas for the adventures from children's

suggestions.

There are many more adventures in the later Oz books. Dorothy

doesnt just get caught in a cyclone, she fell into an earthquake, gets   20

blown off a ship at sea, and almost gets turned into an ornament by

a wicked Nome King. To anyone who thinks there is only one Oz

story, they are in for a delightful surprise.

PART II

Putting a Paper Together

What to Do When You're Stuck

FREEWRITING

For practice in freewriting, use one of these starter sentences and write nonstop for fifteen minutes. Write at least one page. If you get stuck, don't stop writing; instead, write the word *stuck* over and over until a new idea comes to you. Don't stop to think or correct errors. After you finish, go back and proofread and correct what you've written. Over a period of several days, freewrite about each of the seven topics.

1. Right this very minute, I feel . . .

2. I am a person who has strong opinions. One of my opinions is . . .

3. My birthday turned out to be a big . . .

4. The major goal I have to accomplish this month is . . .

5. If I could change just one condition in American life right now, it would be . . .

6. As she walked along the sand, Marlena saw . . .

7. A great new gadget would be . . .

Finding an Organization for Your Essay

Choose one of the following topics, for writing a paper:

- How elderly people are treated in a checkout line in a grocery store
- How one particular person influenced you
- How you size up people from the clothes they wear

Follow these steps:

1. Make a long list of ideas and examples that you could use for your topic.

2. Decide which are your main points and which examples and lesser ideas go under each of them.

3. Now write a single sentence that sums up what you want your paper to say.

4. Think about the connection between the ideas. Then arrange the major ideas in a logical sequence so one will lead gracefully to the next.

How to Make a Paper Longer (and When to Make it Shorter)

Read the following student essay. As you read, mark which details are strong, which details should be cut or condensed, and which details should be expanded or explained.

Primitive Years

The cracked wood on the outside of the shabby three-room schoolhouse had changed from a glossy white to a dingy brown. The front of the school was almost all windows. There was a large window at each room. The window was divided with wood into a lot of small square windows. About five of the windows were nearly always broken. In several months these would be fixed and they would get broken again and again and again. The roof of the schoolhouse was tin and rounded. It reminded me of a barracks. The rooms had gritty-looking hardwood floors like you see in some old general store. Each room was heated by a small coal stove which sat in the middle of the room.

The faculty consisted of three members. Each teacher had two grades in one classroom. My teacher had grades one and two, another teacher had grades three and four, and the other teacher

5

10

115

taught grades five and six and was also the principal. There were   15

about fifteen students in each grade. While the first-grade kids in

our classroom were having arithmetic, we would be working on an

assignment and vice versa. The teachers were very strict and it was

very hard to cut class. It was even harder if you did cut class—on

your bottom!   20

There were no clean indoor restrooms but instead there were two

outside toilets or outhouses.

There wasn't any cafeteria so we had to bring our lunches. We ate

them in the classroom at our desks. The school offered milk for a

small price, like three cents. As a special event on Fridays we could   25

have chocolate milk if we preferred.

Beside our schoolhouse were three seesaws balanced over a long

sawhorse. This was our only playground equipment, but we survived.

Across the creek from the schoolhouse was a little clearing which

we called "the flat." At recess we would go to the other side of the   30

road to the flat and play.

After I finished the second grade, we moved to more modern

civilization. Today, I look back on this and think how primitive this

was, and yet it was only thirty years ago.

Introductions and Conclusions

1. After you've finished the exercise on "Finding an Organization for your Essay" (on p. 114)—or an essay that you've already written—use the following guidelines to write three introductions that you might use.

   - Start with a shocking statement

   - Start with an opinion you wish to oppose

   - Start with a brief story

2. Now write three conclusions, using these guidelines:

   - Sum up your main points

   - Raise further questions or implications

   - Put your ideas in a wider perspective

Paragraphs—Long and Short

Write a well-developed paragraph on one of the
following topics. Be sure to give the specific details and
to stick to *one main idea.* Before starting each topic,
list a number of points you might use; then choose the
best point and write your paragraph to back it up.

1. One strength or weakness that you have

2. One reason you chose your college

3. One statement summing up the style of the room
   you are in right now

4. One good point or one bad point about the
   president of the United States

Transitions

Choose one of the transition words from the following list to complete each sentence below. Be certain that the word you choose makes sense in the context of the sentence. Do not use the same word more than twice.

Consequently	therefore	furthermore	since
in fact	however	eventually	though
nevertheless	above all	for example	finally

Example: *Since* the snake was creeping towards us, we.grabbed a pointed stick.

1. Sleep is necessary to good mental health; _____, too much sleep can sometimes lead to depression.

2. Someone who plays the piano, _____, can add a lot to a party.

3. If you're planning to become a social worker, _____, it's very important to get a master's degree.

4. _____, jasmine oil can also be used as an aid for relaxation.

5. I threatened to call the police. _____, the man gave me back my wallet.

119

6. _____ do not fail to sign your tax return or it will be promptly

returned to you.

7 ._____ UPS will not deliver to a post office box, I will mail

the package parcel post.

8. Mrs. Lacy avoids the sun completely; _____, she carries a

parasol whenever she goes out.

9. _____, France raised $400,000 in order to build the Statue

of Liberty as a gift to America.

10. Hard work usually pays off. _____, Steve Jobs and Steve

Wozniak created the Apple computer in their garage.

Writing Activity

Write a short essay—at least three paragraphs—
about going on a trip. Make a conscious effort to
use appropriate transitions both between
sentences and between paragraphs.

Proofreading

Proofread and correct the following paragraph for typographical and other errors. For example, the first line should be corrected to read:

Millions of people suffer from sleep disorders. Although

Milllions of people suffer from sleep disorders and, although sleep

disorders are not exactly dangerous to one's health, they are

excruciating for those who have them. Some times sleep disorders

can be alleviated by simple methods such as going ot bed later,

drinking less caffeine, or undertaking a mild exercise routine at   5

bedtime. Often, however,the problem doesn't go away so easily and

eventually people who suffer from prolonged sleep disorders will

develop other other health problems, such as constant fatigue,

depression, migraine headaches, and a general breafdown of the

immune system. But thats not all. Sleep disorders are expensive.   10

Sufferers have to pay for treatment out of there own pockets since

most insurance companies do not recognize sleep disorders a a

medical problem, however, since Sleep disorders are   not regarded

as a major health threat, little medical research is being done on

them and therefore no real treatments have been   15 developed. Until medical science, decides that sleep disorders are a

serious problem which must be addressed, millions of people will

continue to walk the floor night after night and suffer alone, without   20

hope.

Proofreading

Proofread and correct the following paragraphs for typographical and other errors.

For example, the first sentence should be corrected to read:

One of the most influential books of the past century is <u>Uncle Tom's Cabin</u>. Although

One of the most influential books of the past century is <u>Uncle

Tom's Cabin</u>. Although some critics consider it sentimental,

overwritten, and contrived, nearlrly everyone acknowledges that the

book did much to inst9gate the American civil War.

Harriet Beecher Stowe, the author the the book, grew up in a   5

home which often offered refuge to runaway slaves who were

escaping on the Underground Railroad into Canada. As a young girl,

Stowe heard many stores directly from the mouths of these slaves

about the horrors they had endured. Her feeelings for these abused

people, coupled with her religious fervor, inspired her to do   10

something to help such a help correct such a wrong.

Stowe's father, brothers, and husband were all ministers who

fought against slavery from the pulpit. When her sister-in law wrote

to her, pleading wih her to "write something that would make a

123

whole nation feel what an accursed thing slavery is," Stowe   15

Resolved that she would do so. She wrote what she considered the

climax of the story—Tom's death—first and then went back and

wrote the beginnig of the story. In 18511, the <u>National Era</u>, an

antislavery paper, began serialization off the book and, when the

book itself was issued in 1852, it sold out within two days. Uncle   20

Tom's Cabin made Stowe an international figure, despised by some

and admired by most.. President Abraham Lincoln honored her and

called her the "lady who wrote the book that made this big war."

Proofreading

Proofread and correct the following paragraphs for typographical and other errors.

For example, the first line should be corrected to read:

If you ever drive through Mississippi, you might be surprised to

If you ever drive through Missippippi, you might be surprised to

see the diversity of landscape. HJeading into the state form

Memphis, Tennesseee, you'll pass Oxford, home of Ole Miss, and

within minutes you'll be driving up and down very thick forested hills

which are covered in kudzu, a rich gree climbing vine that is   5

indestructible.

Leaving the northern part of the state state, you will enter a very

low valley call the Delta—so named because of its similarity to the

delta rgion in Egypt. For many decades, the delta's rich fertile soil

produced outstanding cotten crops and put mississippi on the map   10

as the center of the cotton industry.

After you pass throughf the Delta region, you'll begin a climb again

and with in an hour will arrive at a small ciyt nestled into romantic

bluffs overlooking the Mississippi River. This is Vicksburg, the site of

one of the most decisive battles of the War between the States   15

At this point, you can travel along the Missssippi River down to
Natchez where still stand some of the most magnificient ante-
bellum mansions in the America. From Natchez you can go down
into St Francisville Louisiana, and stop to see several large old
plantations where sugar sane used to grown.   20

However , if you want to take another see the rest of the state,
you can drive west from Vicksburg and in an hour you'll be in
Jackson, the state's capital, and the only city in Mississippi with a
population of over 150,000. From Jackson, you'l;l descend again
through the flat piney lumber region which takes you through   25
Hattiesburg and then on to the Gulf Coast. The coast, situaated
along a twenty-mile beach overlooking the Gulf of mexico, is dotted
with unusual little shrimping towns such as Biloxi, Ocean Springs,
and Bay St. Louis. From the coast, it's only a ninety-minute drive
along eth Gulf of Mexico to America's most exotic city—New Orleans,   30
"The City that Care Forgot"

Proofreading

Proofread and correct the following paragraph for typographical and other errors. For example, the first sentence should be corrected to read:

Every March the emperor penguins, found in Antarctica, gather together in huge masses on the bare ice.

Every march, the emperor penguins, found in Antarctica, gather

together in huge masses on the bare ice. The emperor is the largest

species of penguin. Their weight; is about 65 pounds and they stand

almost four feet tall. They dont build nests. The femal lays a single

large single large egg. She doesn't allow it to lie on the bare ice,but   5

passes it from her large webbed feet to the male's feet. Then all the

males huddle together with the eggs on their feet. A loose fold of

skin covers each egg and keeps it warm. Whilethe males incubate

the eggs in this strange manner, huddling together for warmth in

colonies of up to 5,000, the females head off to sea to feed. The   10

males stand there, braving the coldest weather conditions on earth,

with no food, for up to two months. They survive on their body fat a

lone. Two months later the females return with full bellies; the eggs

hatch and the males, weak and hungry, with half their body weight

gone, waddle to the sea for a well-earned meal.   15

PART III

Meeting Specific Assignments

Format

The page below has numerous errors of format. See how many you can list.

"DREAMS"

by *Lee Dunsworth*

In the Old Testament, Joseph interpreted the dreams of Pharaoh and prophesied the future. Ever since, people have believed that dreams carry secret kno-wledge. For centuries they thought that through dreams they could learn the future. But in 1900, [5] Sigmund Freud turned dream interpretation on its head:he used dreams to understand the past. Most psychotherapists follow Freud's example. However ,some modern neuroanatomists question whether dreams mean anything at all.

[10]

In the Bible, Joseph interpreted dreams on two occasions.

Writing in Class

Choose one of the following topics:

Evaluate someone who has supervised your work—currently or in the past. How well does this person do his or her job?

Compare your home to a home that you have visited which is very different from yours.

Now take fifteen minutes to do the following steps for the topic you have chosen:

1. List as many points as you can for use in an essay. Write a word or phrase for each.

2. Under each point, list examples that would demonstrate or support that point.

3. Decide on two or three main points and the examples or reasons that will make each point clear.

4. Plan the order of these points.

5. Write an introduction to the essay, indicating your main points.

Writing in Class

For practice in writing under pressure, set a timer for
an hour and write an essay on one of the following
topics. Plan for ten minutes; write for forty minutes;
and proofread for ten minutes.

1. Recall an experience—good or bad—which changed
   your life. Describe the experience briefly and then
   explain in detail its influence on you.

2. Think of something you want in your life right
   now, and explain two or three reasons why you
   want it.

3. Argue one side or the other on the use of corporal
   punishment in the home.

Writing About Literature

Here is a poem by Walt Whitman that you can use to practice formulating ideas about literature:

When I Heard the Learn'd Astronomer

When I heard the learn'd astronomer,
When the proofs, the figures, were ranged in columns before me,
When I was shown the charts and diagrams, to add, divide, and
    measure them,
When I sitting heard the astronomer where he lectured with much   5
    applause in the lecture-room,
How soon unaccountable I became tired and sick,
Till rising and gliding out I wander'd off by myself,
In the mystical moist night-air, and from time to time,
Look'd up in perfect silence at the stars.   10

1. Read this poem several times. Read it aloud at least once. Look up any unfamiliar words in the dictionary.

2. Write a brief summary of the poem: in a sentence or two, say what the speaker *did* that night.

3. Locate what you consider the turning point or major shift in the poem. Write a sentence or two explaining the change. List everything you can note prior to the turning point and after the turning point. Include particular choices of words that add to the feeling in each part.

4. Write several sentences explaining the feelings expressed in line 5 and in line 8.

5. Now write a paragraph or two explaining what you think Whitman is pointing out in this poem. Tell which details and choices and which words and phrases make Whitman's ideas clear to you.

135

How to Quote from Your Sources

Here is the "Postscript" from *Rules of Thumb*:

> You do your best work when you take pleasure in a job. You
> write best when you know something about the topic and know
> what you want to stress. So, when you can, write about a topic
> you've lived with and have considered over time. When you *have*
> to write about a topic that seems boring or difficult, get to know   5
> it for a while, until it makes sense to you. Start with what is
> clear to you and you will write well.
>
> Don't quit too soon. Sometimes a few more changes, a little extra
> attention to fine points, a new paragraph written on a separate
> piece of paper will transform an acceptable essay into an essay   10
> that really pleases you. Through the time you spend writing and
> rewriting, you will discover what is most important to say.

1. Write a summary of the "Postscript" in two
   sentences.

2. Use the authors' names (Silverman, Hughes, and
   Wienbroer) in a sentence and give a direct
   quotation from the "Postscript. "

3. Select a word or short phrase that you find
   important and write a sentence telling why.
   Include the word or phrase in your sentence.

4. Respond to the advice in the beginning of the
   second paragraph: "Don't quit too soon. " Write
   two or three sentences based on your own
   experience. Include the quotation within one of
   your sentences.

5. State an opinion about the "Postscript, " followed
   by a quotation from it that illustrates your point of
   view.

Using the College Library and the Internet

Using the public access catalog or computerized catalog, find the answers to the following questions:

1. What is the very first book about the subject "music" listed in the catalog?

2. How many titles by Willa Cather are in your college library?

3. Write down the city of publication, the name of the publisher, and the date of publication of *Working*, by Studs Terkel.

4. Find the call number for *The Prairie Years*, by Carl Sandburg. Go to the stacks and find out whether it is on the shelf. Write down the titles of three books you find closest to its place on the shelf.

Find the answers to the following questions by using the library resources indicated:

5. Using an Internet search engine such as *Altavista* or *Yahoo*, find several articles about wolves in Yellowstone National Park.

6. How did Joe DiMaggio do at bat on 17 June 1941? Use *The New York Times* on microfilm or CD-ROM.

7. When, in 1863, did New Yorkers learn of the fall of Vicksburg, Mississippi, to Union forces? Use *The New York Times Index*.

8. What is the origin of the phrase "rule of thumb"? Consult the Oxford English Dictionary (OED).

9. How many articles about President John F. Kennedy were listed in *The Reader's Guide to Periodical Literature* for 1990?

10. On the Internet, go to www.dejanews.com and submit a query "Why are barns painted red?" Review the answers until you find the most plausible explanation.

Using the College Library and the Internet

Choose one of the following writers for a bibliographical search:

James Baldwin
Flannery O'Connor
Leo Tolstoy
Langston Hughes
Henrik Ibsen
Edith Wharton

1. Write down the years of the writer's birth and death.

2. Write down the titles, publishers, and original years of publication of three books by the writer.

3. Write down the authors, titles, and years of publication of three books about the writer.

4. Write down the authors, titles, names of periodicals, dates, and pages of three articles about the writer.

5. Use either the Humanities Citation Index or CARL.ORG to find an article on one of the authors.

Note: Internet information from Rules of Thumb is regularly updated on the McGraw-Hill Website at www.mhhe.com/socscience/english/compde/ewtitl.mhtml

Writing Research Papers

Make believe that these are some of your notes from research on pizza. You were very careful when taking notes, so only the words surrounded by quotation marks are direct quotations; all other phrases are already paraphrased (in your own words), and you double-checked the page references before you left the library.

Notes

Jeff Smith advises adding just one extra ingredient to tomato sauce and cheese on the dough—for example, tomatoes, or onions, or potatoes "arranged like tiles. " He also says to "drizzle some good olive oil over the whole thing" before baking (487).

Alice Waters makes a pizza with a layer of tomatoes and a layer of shrimp sautéed with scallions; she adds chopped basil and parsley after baking. There is no cheese. She also says to "Try this pizza with a mixture of shellfish: squid tentacles, shrimp, mussels, scallops, and so on" (160).

In *The New York Times International Cookbook*, Craig Claiborne uses anchovy paste all over the dough before he spreads on the tomato sauce. Then he adds grated parmesan and chopped mozzarella. This is his basic recipe, but he omits the anchovy paste when he adds mushrooms or pepperoni (440).

1. First read over the notes to familiarize yourself with the information.

2. Next, put these notes aside and write a paragraph about your own favorite pizza combinations.

3. Now add information from your notes to your paragraph where it will fit in smoothly to support your remark or contrast with what you have

already written. Include at least two direct quotations.

4. Add transition words and phrases, such as *according to, on the other hand, in contrast,* or *similarly.*

5. Add parenthetical citations as needed.

6. Revise your paragraph for logical order, and edit for correctness.

Writing Research Papers

Consult a newspaper, watch television news, or listen to a public radio station (or an all-news network). Take notes on a major event that is reported—such as a sports game, a political announcement, a person's accomplishment, an unusual accident, or a weather story. Be sure to write down at least one phrase, word for word. If you are getting the story on radio or on television, you might want to tape-record the broadcast.

Now write a paragraph which summarizes the event, incorporating the quotation as well as your own comments. Make sure that most of the paragraph is in your own words. Be sure to acknowledge your source.

For example, you might begin:

> Today the city council approved a new recycling law that will affect every citizen in our locality. The mayor announced, "...

Plagiarism

Here is a passage from page 22 of Jack London's account of the San Francisco earthquake of 1906:

> The earthquake shook down in San Francisco hundreds of thousands of dollars worth of walls and chimneys. But the conflagration that followed burned up hundreds of millions of dollars worth of property. There is no estimating within hundreds of millions the actual damage wrought.
> —"The Story of an Eyewitness." *Collier's Weekly* 5 May 1906: 22–23.

Which of the following statements in a term paper must give credit to Jack London in order to avoid plagiarism? Insert quotation marks where necessary.

1. The San Francisco earthquake of 1906 shook down hundreds of thousands of dollars worth of walls and chimneys. The conflagration that followed burned up hundreds of millions of dollars worth of property.

2. The earthquake of 1906 caused hundreds of millions of dollars in damage——both directly (by knocking down structures) and indirectly (by igniting fires) throughout San Francisco.

3. Earthquakes often destroy much of the property and wealth of a city.

4. One account of the San Francisco earthquake estimated that

    hundreds of millions of dollars in property was burned in fires

    caused by the earthquake.

5. The San Francisco earthquake of 1906 knocked down many

    buildings and virtually destroyed the city.

Documentation

PARENTHETICAL CITATION

Here is a sample paragraph from a research paper on pizza. Identify the places where parenthetical citations are needed—that is, where the information is not common knowledge—and mark them.

For example, a parenthetical citation is needed at the end of the first sentence:

> A *New Yorker* cartoon shows two technicians puzzling over a slice of pizza at "The Bureau of Unidentified Pizza Toppings" (   ).

A *New Yorker* cartoon shows two technicians puzzling over a slice of

pizza at "The Bureau of Unidentified Pizza Toppings." Pizza—with all

its various toppings—is one of the most popular food items in

America. In fact, Leonard Lopate reports that over 20,000 acres of

pizza are consumed daily in the United States. From childhood,   5

Americans eat pizza—at birthday parties, after school or sports

events, on their first dates, or while at work on projects. In

particular, pizza is a wise choice for a quick meal when there's a

work deadline: It doesn't require utensils. It's the lowest in calories

of the three most popular fast-food meals, and it is nutritionists' first   10

choice. Pizza is such standard fare during overtime that news

reporters regularly monitor Domino's Pizzas' deliveries to the White

House to see if a presidential announcement is forthcoming.

144

Documentation

PARENTHETICAL CITATION

Assume that you're writing a paper about Hoagy Carmichael and one of your sources is the following paragraph:

> "Star Dust," the most recorded song in the history of American music, was written in 1927 by Hoagy Carmichael, one of the most inventive composers of our day. Within a few years, the song made Carmichael a famous and wealthy man who then gave up a career as a lawyer to become a full-time musician. He went on to write many other classics—such as "Georgia," "The Nearness of You," and "I Get Along Without You Very Well"—but "Stardust" continued to be the most popular. According to Ronny Schiff, Carmichael got "letters from people the world over telling about romances that bloomed because of this song" (28).
> —"The Music of Hoagy Carmichael," by Elaine Hughes, p. 2.

1. Use the author's name in a sentence and give a direct quotation.
2. Write a sentence including a fact from the article.
3. Comment on a word or short phrase the author used.
4. Write a sentence that includes the quotation at the end of the paragraph.
5. Write a sentence that summarizes the main point of the paragraph.

Now add the appropriate parenthetical citation after each sentence. Make sure that the period at the end of each sentence is *after* the closing parenthesis of the citation.

Note: The quotation from Ronny Schiff comes from her book,

Hoagy Carmichael: The Stardust Melodies of. . . . Melville, NY: Belwin, 1983.

Documentation

Here are five sources on pizza. Arrange them in the proper format for a Works Cited page:

1. Jeff Smith wrote The Frugal Gourmet Cooks Three Ancient Cuisines: China, Greece, and Rome. The book was published in New York City by William Morrow and Company, Inc. in 1989.

2. Gourmet magazine gave a recipe for breakfast pizzas in an article called Cuisine Courante: A Spring Brunch in the April 1992 issue. The recipe was on page 219, but the article began on page 128 and continued on various pages throughout the magazine.

3. You interviewed a nutritionist at Meadowbrook Hospital on October 12, 1993. Her name is Dr. Merrillee Turly.

4. On the fifth of April, 1989, Marian Burros in the New York Times compared pizza to focaccia in an article called Savory Marriages Made in California—section C, page one, continuing on page six.

5. Although it's called an "encyclopedia," encyclopedia of European Cooking is actually an edited book and should follow the format for a collection or anthology. The editor is Musia Soper, and the book was published in 1962 in London, England. Spring Books was the publisher; the chapter you used is entitled Italian Cooking, on pages 351–400, written by Dorothy Daly.

A Note to Teachers and Students

The following research paper was written by a freshman at Nassau Community College. We have made necessary changes only in format and documentation, but have left other elements of the paper (logic, organization, style, and use of evidence) as written.

Dirty Times

Jenifer Franceschi-Wood
English 102 FA
Dr. J. R. Silverman
20 Apr. 1998

The American people have an obsession with personal cleanliness; that is why the market for soaps, shampoos, and deodorants is such a good one. No matter what the marketing strategy, such products will always sell. It is not a coincidence that to be clean and well deodorized has become especially desirable since the invention of television. Surely, the last forty or fifty years have been particularly well deodorized because television tells us that being clean and smelling good is desirable.

However, long before television and the mass media, we were a rather dirty bunch of human beings. Today, we take for granted a simple shower or bath with soap and fresh clean water, but in the past keeping clean was not quite so simple. In fact, a clean and deodorized America would be shocked at how seldom our recent ancestors bathed. After all, people didn't always have deodorant and soap, so sometimes plain water had to suffice (plain water that was not clean in and of itself, to say the least). How did they do it? How did they keep clean in medieval times and

in the Victorian Age? They didn't do it very
well, and there is an enormous difference between   25
the personal habits of the rich and the poor back
in those times.

Medieval England was a dangerous place to
live. Frances and Joseph Gies explain that often
the plague was rampant and the person who was   30
lucky enough to live past childhood usually did
not live past the age of forty-five (121).
Medieval peasants rarely if ever bathed, and one
can only wonder if the spread of the plague and
other diseases could have been curbed with the use   35
of a little personal hygiene. Gathering the water
was a great deal of work, so unless it was
specified for cooking, it was not seen as worth
the effort. Gies and Gies tell how a family of
five would take their baths about once a month and   40
usually use the same bath water in succession. A
barrel with the top removed served as a bathing
vessel (93). On a particularly dirty day, say, if
a peasant was covered in mud from working in the
fields, he or she might wash with some plain water   45
to get rid of the grime (Buehr 134). Farm animals
moved freely in and out of the hovels and slept a

matter of a few feet away from the occupants of
the home (Howarth 12). When tidying up the house
it was normal to sweep up animal dung along with
the rest of the dirt on the floor (Gies and Gies
92-93). In towns, some people bathed in public
bath houses called "stews." Stews were well known
for having open nudity between the sexes, and
loose women could frequently be found there (Buehr
134). No wonder there was so much death from
disease.

The rich lived slightly differently in
medieval times. Some of the rich saw baths as
having medicinal qualities. King John washed in
"sweet green herbs" with "five or six sponges to
sit upon" and rinsed with rosewater (Hassall 210).
A contemporary record of these procedures explains
that first one would "boil leaves and herbs" and
then

> Throw them hot into a vessel and put your
> lord over it and let him endure it for a
> while . . . whatever disease, grievance or
> pain ye be vexed with, this medicine will
> surely make you whole. . . . (qtd. in
> Hassall 211)

Apparently the rich had the luxury of using baths
as a cure for ills while the poor used them
sparingly. Still, rich and poor of medieval times
were both afflicted with insects and vermin. It   75
was a rare person who wasn't tormented by lice and
fleas (Collis 10). One can only wonder if this
was because of the lack of real soap rather than
herbs & leaves. Charles Panati reports that soap
dates back to the Phoenicians in 600 BC, but soap   80
production "virtually came to a halt" in the
Middle Ages when the Christian Church forbade
"exposing the flesh, even to bathe" (218).

I'm not sure what happened to "cleanliness is
next to godliness," but even before the Church   85
made modesty more important than getting rid of
dirt, it was clear that the English society didn't
care much for sanitation. The Romans had
introduced the concepts of sewage disposal and
providing a clean water supply for drinking and   90
bathing to England in 43 AD, but Quennell and
Quennell write that "Incredible as it may sound,
no real advance was made for 1400 years; from the
early fifth century until the beginning of the

nineteenth century people were not concerned with
Public Health" (88).

Things did not get much better--at least for
the poor--in Victorian England. Many people think
that the Victorians were prudish about their
bodies and therefore would not allow themselves to
go unwashed. This is only partially true. In
Victorian England the well-to-do were the ones who
kept stately homes and wore fine clothes. The
wealthy had frequent baths and washed daily with
water and soap (Childers 406). A basin was kept
in the bedroom and every morning it was filled
with fresh water to wash with. When the regular
bath was taken, a high-backed tub called a "hip
bath" was placed on the bedroom floor over some
towels to catch any water that splashed out. The
servants carried fresh hot water up in metal cans.
The whole process was laborious, but the
Victorians enjoyed it a great deal. At this point
there were bathrooms, but most wealthy Victorians
preferred to bathe in their rooms. The thought of
sharing bathing facilities with other family
members was repugnant to them (Quennell and
Quennell 71).

The poor were not so lucky in Victorian

England. Since the poorer class accounted for the 120

masses in Victorian England, a great many people

lived in filth and muck. According to E. Royston

Pike, it was normal for two or three families to

live in one room (at a lodging house) and to

urinate and defecate in a pail in the center of 125

the room. Beds were always filthy and full of

worms and insects; people were plagued by lice and

fleas (Pike 298-99). The rooms in the lodging

houses were usually unventilated and waterless.

In 1849, medical reports explained how leaking 130

pipes released contaminated water into the London

reservoir; thus a whole neighborhood could be

contaminated by water containing raw sewage (qtd.

in Pike 306). The 1850 *London News* reported

"routine city flooding when the Thames 'backed up'" 135

(qtd. in Gayman 5). Sometimes water lines would

run through cemeteries and pick up decomposing

animal matter (Pike 281). When the people did

bathe, either the water was black with dirt and

sewage or dripped sparingly from a pipe that a 140

whole village waited in line to use (Pike 79).

Ironically, the lodging houses would not admit

154

people unless they had cleaned their feet (Pike 298).

It is entirely possible that the obsession with cleanliness that we know today stems from the modern knowledge that filth breeds disease. However, I think that this knowledge has evolved into neuroticism. It is not enough to be thin and beautiful or rich and handsome. One must be fresh and clean too. That brings to mind a commercial for a deodorant that I saw just recently. A young woman ruminates on how gross it is when she gets really close and a guy smells bad. I think this is why the producers of that deodorant make so much money; they shame us into buying their product. The poet Galway Kinnell calls advertisers "anti-prostitutes" because they "loathe human flesh for money" (43).

Still, when I walk into a store or a classroom and my nose is invaded by the stench of someone's body odor, I cannot help but wonder if he or she has seen that commercial or passed by the deodorant aisle of the supermarket lately. It occurs to me that television has gotten me too. Perhaps if I were back in Victorian or medieval

145

150

155

160

165

times, I wouldn't care so much about that person's
body odor; but then again, neither of us would
have much choice.

Works Cited

Buehr, Walter. When Towns Had Walls: Life in a

    Medieval English Town. New York: Crowell, 1970.

Childers, Joseph W. "Observation and

    Representation: Mr. Chadwick Writes the Poor."

    Victorian Studies 37 (1994) 405-31.

Collis, Louise. Memoirs of a Medieval Woman: The

    Life and Times of Margery Kempe. New York:

    Perennial, 1983.

Gayman, Mary. "The Badd Olde Days." Cleaner. n.d.

    Cole Pub. [Three Lakes, WI]. 25 Oct. 1997

    <http://klingon.util.utexas.edu/londonsewers/

    londontext3.html>.

Gies, Frances, and Joseph Gies. Life in a Medieval

    Village. New York: Harper, 1990.

Hassall, W. O. Medieval England as Viewed by

    Contemporaries. New York: Harper, 1967.

Howarth, Sarah. Medieval Places. Brookfield, CT:

    Millbrook, 1992.

Kinnell, Galway. "The Dead Shall Be Raised

    Incorrectible." Book of Nightmares. Boston:

    Houghton, 1971. 39-43.

Panati, Charles. Extraordinary Origins of Everyday

    Things. New York: Harper, 1987.

Pike, E. Royston. <u>Golden Times: Human Documents of</u>

<u>the Victorian Age</u>. New York: Praeger, 1967.

Quennell, Marjorie, and C.H.B. Quennell. <u>A History</u>

<u>of Everyday Things in England: The Rise of</u>

<u>Industrialism</u>. London: B.T. Batsford, 1933.

PART IV

Writing with Elegance

Finding Your Voice

Here are some experiments to try, to help you discover the range of your writing voice. Write quickly, with energy; do each exercise for at least five minutes.

1. Complain: Get out your thoughts about something that bothers you.

2. Curse: Tell what you'd like to have happen to someone who makes you mad.

3. Praise: Tell how wonderful someone or something is.

4. Teach: In a helpful, encouraging way, tell a beginner how to do a simple task you do well.

5. Be blunt: Tell it like it is. Say the truth about something most people don't admit.

6. Be sarcastic: Throw scorn on something despicable.

Write a letter to someone important to you. Tell that person exactly what you couldn't say face-to-face, being as honest as you are able. You don't have to send the letter . . . but you might decide to do so.

Adding Details

Here are three general statements. Choose one and write a paragraph using details to illustrate it.

1. You can tell a lot about any family by looking at the inside and outside of the family's refrigerator.

2. Traveling on $20 a day can be a lot of fun.

3. When you're in physical pain, the world becomes a different place.

Write down every detail you can observe about each of the following items. When possible, describe what you can sense about the item using your eyes, ears, nose, fingers, and tongue. Take fifteen to twenty minutes for each, writing as you observe. When you think you have noticed everything, pause and then observe again.

1. An apple

2. The space you are in—a circle of a three-foot radius around you

3. A tree

4. A photograph from a magazine

5. An article of clothing

Recognizing Clichés

1. Make a list of clichés you hear at a social gathering or during one hour of television and rewrite them to make them fresh.

2. Describe a situation in which people use clichés. Tell what each person says and what you think is really going through the mind of one or all of the people involved. Write several pages if you can.

   Some possibilities:

   - a first date
   - being stopped by a police officer
   - consulting a doctor
   - seeing a former friend
   - visiting a sick relative
   - having Thanksgiving dinner

Eliminating Biased Language

Rewrite the following sentences to eliminate bias. For example, examine the following sentence:

> The local underwear factory is hiring seamstresses and machine repairmen.

This sentence can be revised to read:

> The local underwear factory is hiring people to sew and people to repair machines.

1. The automobile repairman should bring his own tools for the training seminar.

2. Attorneys, dentists, and doctors often keep photographs of their wives and children on their desks for clients to admire.

3. The breadwinner of the family should list his income on line 2; the housewife's income, if any, is listed on line 3.

4. The sorority girls did a workmanlike job on their float for the homecoming parade.

5. The policemen don't have too many details about the robbery; the gunman left no clues to his identity, and the victim is still in a coma.

6. John Smith is a noted white writer.

7. Volunteers manned the telephones during the fund drive.

8. This committee represents the entire community—all racial groups, a woman, and both religions.

9. The printing department is overworked: each teacher should have her copying requests submitted two weeks prior to the date needed.

10. Our airline provides top service: all repairmen use the latest aerotechnology, the stewardesses work solely for passenger safety and comfort, and the pilot is always the best in his class.

Trimming Wordiness

The following sentences can be condensed. Cross out excess words and rephrase if necessary. For example, examine the following sentence:

At six p.m. in the evening, the sound of church bells could be heard in the air.

This sentence can be trimmed to read:

At six in the evening, the church bells rang.

1. Lily Tomlin is a woman who has a zany and offbeat personality.

2. At that point in time, the officers proceeded to make an official arrest of the bank president.

3. John Kennedy was a person who enjoyed a good, humorous joke.

4. The reason why Vice-President Agnew left office was because of taking a bribe secretly.

5. In my opinion I think that Frank was not always pleasant to be around.

6. A slow drip could be noticed coming from the pipe under the sink in the kitchen.

7. Hot, spicy jalapeño chili peppers are something that Mexican chefs really love to use all the time.

8. Renata Bellini quit singing due to the fact that her audiences were not listening most of the time.

9. It really was so hot in Timbuktu during the month of January.

10. The bedspread's colorfast, engineered design provides refreshingly different, interesting, decorative accents for your bedroom.

Trimming Wordiness

In the following sentences, replace or delete most of the *being* verbs *(am, is, are, was, were, be, being, been).* You may have to rearrange the wording. For example, examine the following sentence:

> The basket of figs was left on the table by Mrs. Anchor, who is my cousin.

This sentence can be trimmed to read:

> Mrs. Anchor, my cousin, left the basket of figs on the table.

1. The rats were observed to be disoriented after the test substance was eaten by them.

2. Sophie is a grump who always makes life miserable for all of her friends; even when she is out of sight, the memory of her sour disposition is present.

3. The accident had been reported by the police, and within minutes the firetrucks were there.

4. The electronic percussion was added by the sound engineer after the keyboard solo had been recorded.

5. Our former suppliers have been changed because it was found that the contamination was caused by their negligence.

6. We were having a hard time loading the watermelons onto the truck because they were constantly rolling off and were breaking when they hit the concrete.

7. A man who was wearing a gray raincoat was being watched by the security officer because she suspected that he was shoplifting.

8. When life is tough, I am one of those people who are tougher.

9. In *The Kiss of the Spider Woman*, Sonia Braga's hair was long, her nails were long, and her skirt was tight; she was very glamorous.

10. It was sad to me when I was told that the old railroad depot was scheduled to be demolished.

Writing Activity

Write a paragraph relating an action you observed or took part in—for instance, a street fight, a roller coaster ride, or a dance. After you've written the paragraph, go back and circle all the *being* verbs (*am, is, are, was, were, be, being, been*). Then rewrite the paragraph eliminating all *being* verbs and replacing them with action verbs. In accomplishing the task, you may need to rewrite, combine, or eliminate some of your original sentences.

Varying Your Sentences

COMBINE CHOPPY SENTENCES

A good way to practice varying your sentences is by sentence-combining—taking five or six short sentences and reducing them to two or three. For example, examine the following group of short sentences:

1. Kevin started driving when he was only ten.
2. He drove a truck on his grandfather's farm in west Texas.
3. His grandfather could watch him on the farm roads.
4. He had the farm roads all to himself.
5. He learned to drive in a safe way.

Here are two ways that these sentences could be combined without losing any information:

> Kevin started driving a truck on his grandfather's farm in west Texas when he was only ten. Because he had the farm roads all to himself and his grandfather could keep an eye on him, he learned to drive in a safe way.

> When Kevin was only ten, he learned in safety to drive on his grandfather's farm in west Texas. Under his grandfather's eye, Kevin drove a truck on farm roads that he had to himself.

Varying Your Sentences

COMBINE CHOPPY SENTENCES

Combine the following sentences, retaining the facts. Each group should be reduced to two or three sentences. Be sure to use subordination or transition to show the relationship between the ideas. For an example, see p. 171.

1. The party was not a success.

2. The party ended in a major fight.

3. There was not enough food.

4. The hosts ran out of soda.

5. The guests hated the music.

6. The guest of honor left early.

1. They opened the menus.

2. They started looking at the vegetable dishes.

3. All of the people at the table were vegetarians.

4. They each ordered something different.

5. They decided to share.

Varying Your Sentences

COMBINE CHOPPY SENTENCES

Combine the following sentences, retaining the facts. Each group should be reduced to one or two sentences. Be sure to use subordination or transition to show the relationship between the ideas. For an example, see p. 171.

1. Bicycling in the city keeps you alert.

2. You have to concentrate on what is ahead of you.

3. You have to concentrate on what is behind you.

4. You have to concentrate on what is to either side of you.

5. Bicycling can be dangerous.

6. Bicycling also can be fun.

1. Marinate the chicken in a mixture of lemon, olive oil, and parsley.

2. Refrigerate the chicken in the mixture for two hours.

3. Grill it on a medium fire.

4. Brush it from time to time with the marinade.

5. Cook it for fifteen minutes.

6. Turn it over.

7. Cook it for fifteen minutes more.

Varying Your Sentences

COMBINE CHOPPY SENTENCES

Combine the following sentences, retaining the facts. Each group should be reduced to one or two sentences. Be sure to use subordination or transition to show the relationship between the ideas. For an example, see p. 171.

1. He put the blank tape into the machine.
2. He rewound it.
3. The TV was on channel three.
4. The VCR was turned to "TV/VIDEO."
5. He pushed several buttons.
6. He unplugged the TV.
7. He went to the movies.

1. The platypus is a furry animal.
2. It lays eggs.
3. It has a bill like a duck.
4. It has a flat tail like a beaver.
5. It lives in and near water.
6. It has webbed feet.

Varying Your Sentences

STRONG ENDINGS

Rewrite the following sentences to give them stronger endings. For example, examine the parts of the following sentence:

> Officer Langdon burned her uniform after witnessing the depth of the department's corruption.

The sentence can be made stronger by writing:

> After witnessing the depth of the department's corruption, Officer Langdon burned her uniform.

1. Macon ended up with a broken leg when Evan's dog tripped him.

2. It poured for six days after two months of hot, dry weather, brown grass, and water rationing.

3. For the first time, she let him kiss her when the lights went out and the elevator got stuck between the fifth and sixth floors.

4. The state legislators were scandalized to learn about the retirement package—which totaled $900,000—of the school superintendent.

5. They sapped all the energy out of the writing in order to make the document politically correct.

6. The first volley of fireworks exploded in green and yellow showers as the crowd gazed towards the sky.

7. At last the mugger was arrested after attacking five people in one day and injuring two of them.

8. Nell Ann always comes through for us when the chips are down.

9. The audience cheered when one of the sisters threw flour on her face, jumped on the table, and danced a jig during the first act of *Dancing at Lughnasa*.

10. His hair was pulled back in a ponytail to reveal a new gold earring, which was the first thing that his mother noticed.

Varying Your Sentences

IMITATE GOOD WRITERS

Novels often begin with direct and memorable writing. Below you'll find the openings of four novels with a variety of styles. Use the structures of these sentences to write your own beginnings, matching your sentence rhythms to the author's but using your own subject matter.

For example, here is one imitation of Kafka's first sentence:

> As she drove down the dirt road through the woods to her house that night, she noticed, hidden among the trees, a black pickup truck.

> As Gregor Samsa awoke one morning from uneasy dreams, he found himself transformed in his bed into a gigantic insect.
> —Franz Kafka, *The Metamorphosis*

> I had the story, bit by bit, from various people, and, as generally happens in such cases, each time it was a different story.
> —Edith Wharton, *Ethan Frome*

> In that place, where they tore the nightshade and blackberry patches from their roots to make room for the Medallion City Golf Course, there was once a neighborhood. It stood in the hills above the valley town of Medallion and spread all the way to the river. It is called the suburbs now, but when black people lived there it was called the Bottom.
> —Toni Morrison, *Sula*

> From beyond the screen of bushes which surrounded the spring, Popeye watched the man drinking. A faint path led from the road to the spring. Popeye watched the man—a tall, thin man, hatless, in worn gray flannel trousers and carrying a tweed coat over his arm—emerge from the path and kneel to drink from the spring.
> —William Faulkner, *Sanctuary*

177